About the Cover:

Ocean Sunset. This photo is in the Public Domain.

Refresher 1
Third Edition

by Gordon L. Ziegler

Refresher 1
Third Edition

Not copyrighted. In Public Domain.

Freely you have received, freely give.

Last revised September 1, 2016.

Printed in USA by CreateSpace.com

Author

Gordon L. Ziegler

4401 37th Ave SE Unit 17

Lacey, WA 98503-3576

ben_ent100@msn.com

Preface

The sun is in the latter stages of a nova (getting brighter and hotter and hotter to peak much brighter than currently, before getting less and less bright and going out in darkness.), threatening the lives of billions of people with extinction unless somehow the solar process can be mitigated. The author discovered a new relatively inexpensive technology to tame the sun and rejuvenate the entire earth with health and pain free eternal life. That technology is derived in this book.

The Fukushima Japan nuclear power plant disaster is not getting better and going away with time! It is getting worse and worse with time, with potentially so much radioactive contamination to doom earth's entire populations with extinction from radiation induced cancer! This book explains how to completely reverse this problem too! For years we have thought this would take $130 million through rf cavities, klystrons, cooling towers and linear accelerators. But now we learn we may be able to do the construction for five or ten million dollars through laser particle accelerator technology. Japan has already spent over $58 billion on this disaster. What is $10 million in comparison? What if it didn't work? Wouldn't it have been worth it to find out whether or not it would work?

Potentially, the Refresher is the most important invention in human history. If it works as calculated, it would reverse all aging, disease, and decay processes in humans, animals, vegetation, and minerals to young adulthood, and there preserve them. It could also stabilize and save the sun from safety on earth. While designed to be tested in an area as small as five acres, it is designed to be capable of expansion of the active field in increments, as governments permit, until the whole earth is covered simultaneously by the field from one small machine system, and even the sun is saved from its nova by that one Refresher on earth. One Refresher machine could restore the whole earth to Edenic tion and save the sun!

This book will describe the restoration that machine could give, summarize the theory of operation, give the description and specifications of the Refresher, present the builders who could design and build it, with some of the costs and timelines. This book will also present part of a Draft Environmental Impact Statement for the creation and operation of the Refresher.

The Refresher, however, is not alone for good health. It could enable the operation of another important machine—a radioactive waste-free power reactor. The reactor would cost about $5 million, compared to a new nuclear power plant costing $6 billion. And though costing so much less than a new nuclear power plant, the radioactive waste-free power plant (Clean Energy Source or Electrino Fusion Power Plant (EFP Power)) would produce 150% as much electricity generated—1880 MW maximum (or less if desired). It would be 1,000 times as energy efficient as a nuclear reactor. It would hardly need any fuel, and would not need Uranium, Plutonium, or Hydrogen for fuel. It could use any matter for annihilation fuel, but it could easiest use brass or copper. And yet it is totally free of radioactive wastes. The world desperately needs this new energy source, but it is cost effective and safe only when built in the active circle footprint of the Refresher. The closing chapters of this book will recap the derivation of these things as it derives the features of the radioactive waste-free power generating reactor.

There is no working prototype yet. We have only the theory so far. What evidences are there so far, that could test the theory before we invest in it? The Standard Model of Physics requires 61 different elementary particles to construct all known light, matter, but not gravitons. By contrast, the author's Grand

Unification Theory (GUT) and Theory Of Everything (TOE) can construct all matter, antimatter, light, neutrinos and gravitons with different ionization states, fusion states and combination states of only one particle. What can be more parsimonious than that? The Quark Model has seven different formulas for eight different particles—the first and the last are the same. But the masses are not the same! The Quark Model does not have a unique structure formula for each elementary particle, but the author's model of physics does. Also it is not known how to calculate the myriad unique masses of elementary particles through the Standard Model of Physics, but it is possible through Gordon's particle theory to derive every particle mass from first principles. He has already done so for leptons, neutrinos, gravitons and a few baryons. That will be reported in this Third Edition of *Refresher 1*. The author is now working on two new volumes, *Predicting the Mesons* and *Predicting the Baryons*. The fit of the calculated to measured masses so far is two to four place accuracy for each particle. That is a great tribute to the precision of the measurements of the masses so far. Some of these are high mass baryons. The accuracy of measured masses to calculated masses is about the same for high or low masses.

CONTENTS

Chapter	Page
Preface	4
Contents	7
1. Summary of the Author's Grand Unification Theory	9
2. Impossible With Einstein's Special Theory of Relativity	12
3. Impossible With Spinning Point Charges	21
4. Impossible With the Quark Hypothesis	23
5. The Miraculous Effects of the Refresher	25
6. Reversing the Sun	27
7. The Second Law of Thermodynamics	28
8. The Theory	37
9. Is it Safe?	48
10. Should We Do It?	49

Chapter	Page
11. Clean Energy Sources	50
12. Clean Energy Theory	52
13. Project Description	67
14. Refresher 1 Design Specifications	73
15. Trade Secrets	75
16. Trouble Shooting	85
17. Draft Environmental Impact Statement	88
18. Tests for the New Model of Physics	97

Chapter 1

Summary of the Author's Grand Unification Theory

Gordon's Grand Unification Theory (GUT) and Theory Of Everything (TOE) starts with an aether model of Special and General Theory of Relativity. Einstein tried to explain everything with an aether-less theory of relativity. But that leads to contradictions and absurdities. In 1904, one year before Einstein published his Special Theory of Relativity (SR), Hendrik Antoon Lorentz originated an aether theory of relativity (LR). His theory had a preferred reference frame and a universal time. Einstein tried to prove that one uniformly moving reference frame was as good as any other in explaining any situation. But that led to the Twin Paradox and necessary "time slippage."[1] There is no Twin Paradox or time slippage in the Lorentzian Relativity (LR).[2] There is more power in LR than SR to explain gravity and inertia.[3] Einstein had to have warped empty space to explain gravity (that is hard to wrap your mind around). But an aether composed of a sea of gravitons easily explains gravity and inertia,[3] and gives something to warp in space. Einstein's General Theory of Relativity depends on the Schwarzschild Line Element, which depends on the perfect spherical symmetry of the gravitational body like the sun. But the spinning sun is not perfectly spherically symmetric. The diameter of the sun pole to pole is slightly shorter than the diameter at the equator. The famous Schwarzschild Line Element used by Einstein and the author is but the first order solution, where there should be higher order terms to account for the non spherical symmetry of the problem. This might account for Einstein's unaccountable error in the precession of the orbit of Mercury and other tests of General Relativity.

Einstein tried for thirty years to obtain a Unified Field Theory, but failed. The author, however, has had a Unified Field Theory since 1994.[4] The means of uniting the forces is uniting different pairs of forces, so all the forces are united. By utilizing the parsimony postulate to the ultimate, it is possible to derive the major forces and systems in the Universe with but one other postulate.[5]

The Unified Field Theory can be derived using seven postulates.[6] A Unified Particle Theory can be derived with twelve postulates (where the Parsimony Postulate is in common in both sets of postulates).[7] In the Unified Particle Theory, the major building block of the Universe, the proton, has seven sub-particle fractons: four fourth particles (quartons), two half particles (semions), and one whole particle (a uniton). The other nuclear building block, the neutron, has three sub-particle fractons: two half particles (semions), and one whole particle (a uniton). The electron has two half particles (semions). Antimatter has just the opposite charges for everything.

This is contrary to Murray Gel Mann's Quark Hypothesis. The author disproves the Quark Hypothesis and establishes the Electrino Hypothesis in its place.[8]

Einstein's Theories of Relativity, P.A.M. Dirac's spinning point charge electrons, and Murray Gel Mann's Quark Hypothesis are major pillars in the Standard Model of Physics. But they are all wrong! The Standard Model of Physics is in error! The Standard Model of Physics needs to be corrected. The author will take time out here in the development of this presentation to quote his own chapters on these points in the Standard Model of Physics.

[1]Tom Van Flandern, Univ. of Maryland and Meta Research, Open Questions in Relativistic Physics, edited by Franco Selleri (Montreal: Apeiron, 1998), pp. 81-90, as quoted by "What the Global Positioning System Tells Us about Relativity, Meta Research, http://www.metaresearch.org/cosmology/gps-relativity.asp

[2]*Ibid.*

[3]Gordon L. Ziegler, *Electrino Physics*, Draft 2, amazon.com, Chapter 5.

[4]*Ibid*, Chapter 7; Gordon L. Ziegler, *Formulating the Universe* (unpublished book in the author's possession), Chapter 7 and copyright page.

[5]*Ibid*, Chapter 8.

[6]*Ibid*, Chapter 3, 4, and 6; Gordon L. Ziegler, *The Higgs et. al. Universe*, amazon.com, Chapter 11.

[7]Gordon L. Ziegler, *The Higgs et. al. Universe*, Chapter 11.

[8]*Electrino Physics*, Draft 2, *op. cit.*, pp. 183-185.

Chapter 2

Impossible with Einstein's Special Theory of Relativity

The Theory of the Refresher is impossible to derive with Einstein's Special Theory of Relativity. Also it is impossible to get a government or private grant to build a Refresher unless the reviewers are thoroughly un-deceived from Einstein's Special Theory of Relativity. And that is impossible to do in a 25 page grant application. It is necessary to take time out to make a grand synthesis of physics like this tome to break down the barricades to scientific progress by science falsely so called.

There are several faults with Einstein's Special Theory of Relativity: He tried to do everything without an aether. It is impossible to have a Universe that does not either blow up or collapse on itself without an aether. But with an aether, it is simple—just spin the whole Universe. But that is impossible to define without an aether. Scientists still do not have an acceptable model of gravity and inertia. It was easy for me 34 years ago with an aether (see *Electrino Physics,* Draft 2, Chapter 5). That work even united gravity and inertia in one formula—the beginning of a Unified Field Theory, which Einstein tried for 30 years to achieve and could not achieve, but which I achieved in 1994 or earlier[1]. (See *Electrino Physics*, Draft 2, Chapter 7, "Uniting the Forces".)

In his Special Theory of Relativity, Einstein said it was impossible for anything to go faster than the speed of light. But recently neutrinos were clocked going faster than the speed of light.[2-4] Some have disputed those results, attributing them to various experimental errors. But it is impossible to unite all the forces in a Unified Field Theory without some particles traveling faster than the speed of light.[5]

There are several other difficulties with other aspects of Einstein's Special Theory of Relativity such as time slippage, reversing the real and the apparent, and angle of measurement.

Special Relativity's Weaknesses

Except for popularity and preeminence in scientific publications, Einstein's aetherless theory of relativity has not done quite as well as Lorentz's aether model of relativity. Without any preferred reference frame, Einstein's theory has not accounted for the experimental data as simply as Lorentz's theory with a preferred reference frame. Scientists can argue that Einstein's theory can account for all the data. But other scientists can also argue that it cannot account for all the data parsimoniously. The parsimony principle is an accepted principle of physics. It is a postulate in *Electrino Physics*, Draft 2 in Chapter 6. This principle should have weight in this contest.

Another area of difference between SR and LR is the twin or clock paradox. In the twin paradox, one of two twins rockets away from earth a distance into space, turns around, then rockets back. To the traveling twin, the stay at home twin appears to make a round trip in the opposite direction. Which one will age more than the other, or will they age alike? Will each twin claim that the other aged less because of time dilation? Or will there be a difference in the aging? In SR there is a twin or clock paradox.

How does the resolution of the "twins paradox" compare in LR and SR?

In LR, the answer is simple: The Earth frame at the outset and the dominant local gravity field in general constitutes a preferred frame. So the high-speed traveler always comes back younger, and there is no true reciprocity of perspective for his or other frames.

In SR, the answer is not so simple; yet an explanation exists. The reciprocity of frames required by SR when Einstein assumed that all inertial frames were equivalent introduces a second affect on "time" in nature that is not reflected in clock rates alone. We might call this effect "time slippage" so we can discuss it. Time slippage represents the difference in time for any remote event as judged by observers (even momentarily coincident ones) in different inertial frames.[6]

The author of the above quote goes on to give several numerical examples of time slippage for the traveling twin when he turns around, in order that complete reciprocity of frames may be maintained in Einstein's Special Theory of Relativity. Lorentzian Relativity doesn't need any time slippage. It has a preferred frame. Perhaps scientists should make one more unmanned lunar orbit mission with atomic clocks and return of the module to earth for comparison of the atomic clocks with ground clocks, to see if there is any time slippage, or to see if the clocks are as in LR with a preferred reference frame.

The problem is caused in Einstein's SR because he did not believe there was any way of determining an absolute space in special relativity. He postulated that one uniformly moving reference frame was as good as another. The theory of relativity was thought to be consistent no matter what the rest velocity was assumed to be. Therefore twin A should experience length contraction and time dilation as computed by twin B if twin B assumes he is at rest during his flight.

Some physicists try to harmonize the clock paradox through general relativity, as though the accelerations of twin B affect the positions of his clock. "Cyclotron experiments have shown that, even at accelerations of 10^{19} g (g = acceleration of gravity at the Earth's surface), clock rates are unaffected. Only speed affects clock rates, but not acceleration per se."[7]

The clock paradox can be put in better perspective by studying triplets instead of twins. Let triplet X stay at home on planet earth. Let triplet Y rocket out in space similar to twin B, and let triplet Z rocket out in space in the opposite direction as triplet Y. After a period of coasting, let triplets Y and Z simultaneously decelerate and rocket back to earth for a close-encounter fly by the earth and each other. Let X, Y, and Z compare their clocks. Then Y and Z should continue to coast for awhile, then simultaneously decelerate and rocket back to earth for another close-encounter and fly by. Let them again compare their clocks. Finally let Y and Z continue coasting for awhile, simultaneously decelerating and accelerating back to planet earth, where they land and compare their clocks with triplet X.

Who should have slower clocks? X, Y, or Z? Z should expect Y to have time dilation relative to him according to Einstein's theory. Also Y should expect Z to have time dilation relative to him. One cannot appeal to the accelerations and general relativity to harmonize this contradiction, for the accelerations are symmetrical. Triplet X would expect Y and Z to be time dilated equally.

None of this is a problem in LR, which has a preferred reference frame.

Two Theories of Relativity

In 1904, one year before Einstein published his Special Theory of Relativity, Hendrik Antoon Lorentz originated a theory of relativity.

We must make a distinction between Einstein SR and Lorentzian Relativity (LR). Both Lorentz in 1904 and Einstein in 1905 chose to adopt the principle of relativity discussed by Poincare in 1899, which apparently originated some years earlier in the 19th century. Lorentz also popularized the famous transformations that bear his name, later used by Einstein. However, Lorentz's relativity theory assumed an aether, a preferred frame, and a universal time. Einstein did away with the need for these. But it is important to realize that none of the 11 independent experiments said to confirm the validity of SR experimentally distinguish it from LR—at least not in Einstein's favor. However, the issue of the need for a preferred frame in nature is, charitably, not yet settled. Certainly, experts do not yet agree on its resolution. But of those who have compared both LR and SR to the experiments, most seem convinced that LR more easily explains the behavior of nature.[8]

In Einstein's Special Theory of Relativity, relative mass increase, time dilation, and length contraction are only *apparent*—observational phenomena due to rotation in complex space-time. In quasi-relativity, mass increase, time dilation, and length contraction are *real*—caused by motion relative to an aether. The resultant constancy of the speed of light in quasi-relativity is only *apparent*, not *real* as in Einstein's special relativity. Quasi-relativity reverses what is considered *real* or *actual* lengths and times and what are *apparent* lengths and times. The mass, clock speed, and length of a moving observer as seen by himself only appear to be the rest mass, rest clock speed, and rest length. They are not actually rest quantities, and invariant, as in Einstein's theory.

Transformations in quasi-relativity are different than in special relativity. To transform position, time, momentum, or energy (mass) to another frame in quasi-relativity, one must first make an inverse transformation to the aether rest frame according to the direction and magnitude of the observer's velocity relative to the aether, then make an appropriate rotation of the x axis to the direction of the velocity of the target frame relative to the aether, then make a transformation according to the magnitude of the velocity of the target relative to the aether. Every correct transformation should be a triple transformation like this. Such a triple transformation [1) an inverse transformation of increased clock speed, decreased mass, and length expansion when going from the observer to the aether rest frame; 2) an appropriate rotation; and 3) a transformation of decreased clock speed, increased mass, and length contraction when going from the aether rest frame to the other arbitrary frame, where all transformations go by the clock in the aether frame] is not in general equal to either a single transformation or a single inverse transformation. Here is where Einstein gets into trouble with the clock paradox. Not every uniformly moving frame is equivalent. Relative velocities between observer and target can vary from 0 to ± 2V, with time and energy (mass) remaining constant.

The author derived *Relativity in an Ether*[aether] in 1977 when he did not know of Lorentz' work or anyone else's theory of relativity except Einstein's. Today derivations of relativity are a dime a dozen. Thousands of scientists dissent from Einstein's theory of relativity. But their derivations require ad hoc discontinuous hypotheses to reverse the natural relativistic effects in the Galilean transformations. The author's derivations are still the best, employing Newton's third law as a postulate for a smooth continuous derivation from the Galilean transformations. See *Electrino Physics*, Draft 2, Chapters 1-8 and Problem Solutions, Chapters 1-8 (relativity derivations, start of a theory of gravity and inertia, Unified Field Theory, beginning of a Unified Particle model, and Unified Universe derivations). Please observe that the first chapter noted may seem to be boring irrelevant scientific lore.

But it is necessary to establish a natural discovery pattern for our abstract reasoning and pattern recognition to tackle the relativity mystery.

Another difference of the GUT to the Standard Model is that charged sub-particles of like fusion states can fuse to particles of higher fusion states.[9] The secret of why that should be is that when sub-particles orbit or travel faster than the speed of light in the relativistic frame relative to the baseline non-relativistic frame, their radii become imaginary because of the relativistic length contraction formula. The strong electric force equation for these super luminal sub-particles has two such imaginary radii multiplied together. That makes an additional minus sign in the force equation, which makes like charges attract. When two bound sub-particles of a positron collide with 1880 MeV energy or more with two other bound sub-particles of another positron with like oriented spins in the Center of Mass Frame, the four sub-particles are attracted into the same orbit. Then one sub-particle from one positron is more attracted to one sub-particle from the other positron than to any other sub-particle because of closer proximity. These are like charges, and here like charges attract because they travel faster than light. The two sub-particles are attracted by the electric strong force. Nothing stops them from fusing. The other two positron sub-particles fuse also. Four sub-particles fuse down to two particles each with twice the charge as the charge of one of the four sub-particles. The four sub-particles are ½ e charges. The two fused particles are 1 e charges each, but though they are numerically whole particles, they cannot exist alone. That is why on creation they scavenge from the graviton sea the necessary sub-particles to become protons or neutrons. But when the positron four ½ e sub-particles fuse to the two 1 e particles, they switch from antimatter to matter. The fusion of sub-particles in positrons results in the generation of solely positive order energy (quantum mechanical energy in the creation of particles). This phenomenon is theorized to reverse the order to disorder arrow in the second law of thermodynamics [because it is positive order energy as opposed to negative order energy which surrounds us and which determines the current order to disorder arrow direction and the direction of reactions].[10]

---

[1]Gordon L. Ziegler, *Electrino Physics*, Draft 2, Chapter 7; Gordon L. Ziegler, *Formulating the Universe* (unpublished book in the author's possession), Chapter 7 and copyright page.

[2]Frank Jordans and Seth Borenstein, "Neutrinos clocked moving at faster-than-light speed" (Associated Press): http://www.msnbc.msn.com/id/4462927/ns/technology_and_science-science/?gt1=43001.

[3]Robert Evans, "Particles found to break speed of light" (Reuters): http://www.reuters.com/article/2011/09/22/us-science-light-idUSTRE78L4FH20110922.

[4]"Do neutrinos move faster than the speed of light?"—physicsworld.com: http://physicsworld.com/cws/article/news/47283.

[5]*Electrino Physics*, Draft 2, *op. cit.*, Chapter 7.

[6]Tom Van Flandern, Univ. of Maryland and Meta Research, Open Questions in Relativistic Physics, edited by Franco Selleri (Montreal: Apeiron, 1998), pp. 81-90, as quoted by "What the Global Positioning System Tells Us About Relativity, Meta Research, http://www.metaresearch.org/cosmology/gps-relativity.asp.

[7]C. MØLLER, *The Theory of Relativity* (Oxford: Clarendon Press, date unknown to author).

[8]Tom Van Flandern, *op. cit.*

[9]*Electrino Physics*, Draft 2, *op. cit.* Chapter 12.
[10]*Ibid.*, Chapter 16.

Chapter 3

Impossible With Electrons as Spinning Point Charges

Post-Modern Electrons

By Gordon L. Ziegler

Abstract

The modern electron theory was put forth by P. A. M. Dirac in 1928—namely that electrons were spinning point charges. Dirac was satisfied with only a first order approximation fit with measurements—which has been adequate for all the modern era. But very high energy electron collisions in the post-modern era require more precision than that. This paper will identify what was wrong with the spinning point charge theory and how to correct it.

Electrons cannot be spinning point charges in the absolute, because a point charge in the absolute has an infinite mass, and an electron has a small mass. The concept of particle spin models after $mrc = n\hbar$. We can measure the particle mass m and the radius r, but how do we measure the rate of the rotation of the surface of the charge? We could shine a light of some color on the edge of the spinning charge to see what bounces back to determine the speed of rotation. But notice this works only if the charge surface is bumpy or lumpy. If the charge surface is smooth symmetric, the light will just go by without reflecting. This is at the elemental level. Therefore we can take from this a far reaching postulate—a smooth symmetric charge distribution cannot have detectable spin.

It can have spin, but not detectable spin. Common reactions in particle physics model after detectable spin.

A spinning point charge is smooth symmetric. Therefore, if there would be such a thing, it should have no detectable spin. But electrons have detectable spin. Therefore electrons should not be spinning point charges. They should be bumpy or lumpy. The author's model for the structure of the electrons is two half charges orbiting each other at the speed of light. Such a system is lumpy, and it spins. The orbit of the half charges at the speed of light means the half charges are bound in a speed of light barrier. They cannot be blasted apart. They act as one particle. This fact negates the common objection that a two particle electron should be capable of being blasted apart, but electrons always act as single particles.

Various references on spinning point charge electrons and P. A. M. Dirac authorship of the idea are in [1-4]. That Dirac was satisfied with only a first order fit to the measurements is referenced in [4].

---

[1] geocalc.clas.asu.edu/html/GAinQM.html.

[2] en.wikipedia.org/wiwi/Spin_(physics).

[3] en.wikipedia.org/wiki/Electron_optics.

[4] The Quantum Theory of the Electron, P. A. M. Dirac, Proceedings of the Royal Society of London. Series A, Containing Papers of a Mathematical and Physical Character is currently published by The Royal Society.

Chapter 4

Impossible With the Quark Hypothesis

1. How many is too many?

There is a law in science (parsimony principle) that things should be made of as few different kinds of elementary particles as possible. Old science (the Standard Model of Physics with the Quark Hypothesis) requires 61 different kinds of elementary particles to put together light, matter, but **not** gravitons. New science (the Electrino Fusion Model of Physics with the Electrino Hypothesis) requires only one kind of elementary particle to make light, matter, **and** gravitons. Which do you think has the better science—old science or new science?

2. Matching spins and charges

Old science goes by the Quark Hypothesis, which has ⅔ charge and ⅓ charge for the smallest charges to make up everything. They try to match these irregular charges with ½ spins. It is not a very good match. Maybe that is why it takes old science 61 particles to make up light and matter.

New science goes by the Electrino Hypothesis, which has 0 spins for each of the electrinos (1 charge, ½ charge, ¼ charge, and ⅛ charge) and ½ spin for their minimum detectible orbital spins. Doesn't new science have a better match of spins and charges than old science? Maybe that is one reason why new science can make everything out of only one particle.

Explaining Things No Other Theory Can

Electrino Fusion Model versus the Standard Model

Parsimony Principle: It takes the Standard Model of Physics 61 different elementary particles to compose light, matter, but **not** gravitons. It takes the Electrino Fusion Model of Physics only one particle to compose light, matter, **and** gravitons.

Uniqueness: The Standard Model has seven structure formulae for eight particles—the first and the last are the same. But the masses are not the same! The Standard Model does not have a unique structure formula for each particle. The Electrino Fusion Model does.

You cannot make gravitons out of quarks, but you can make them out of electrinos. The Electrino Fusion Model solves for four different gravitons on the lowest chonomic level.

The Standard Model has g/2 factors for two particles. The Electrino Fusion Model has g/2 factors for 28 particles derived from first principles.

The Standard Model cannot derive the masses of light particles from first principles. Those must be input in their model. The Electrino Fusion Model has derived the masses of 22 light and heavy particles to two to four place accuracy from first principles. We are working on the rest. We have a book published explaining the method of calculating the masses of the 22 particles from first principles—see *Advanced Electrino Physics* Draft 2 (orderable through amazon.com or barnsandnoble.com.). See also *The Higgs et. al. Universe by Gordon L.* Ziegler from amazon.com.

Chapter 5

Miraculous Effects of the Refresher

Reverse aging for adults

The simplest effect of the Refresher to understand is reversing adult aging. Old people can be made young adults again in the active footprint of the Refresher. This effect for positron anti-semion fusion does not really back up time or the clock. It merely reverses the order to disorder arrow in adults. It saturates at the maximum state of order—which is young adulthood. It reverses adult aging at a rate of about 1836 times as fast as the rate the original adult aging occurred. A century of aging can be reversed in just over 19.89 days.

Resurrections from the dead

The reverse aging occurs also for bodily remains—re-assembling dust and bones into living beings again. All the dead of all ages of earth's history would be resurrected in under 3.5 years of Refresher machine time, starting with those who died most recently.

Backing diseases out of existence

In the process of reverse aging, diseases would be backed out of existence. This would work also for difficult diseases like HIV AIDS, cancer, and cystic fibrosis.

Reversing all decay

Spoiled fruit would un-spoil in the active footprint of the Refresher. Fresh fruit would stay at the maximum state of order for fruit forever—fresh picked fruit. And this would be without refrigeration. This would amount to a new kind of food preservation without canning or freezing.

This process would un-decay everything in the Refresher footprint, not just fruit. And the footprint could be enlarged to cover the entire earth and even the sun!

Reversing pollution out of existence

In the Refresher footprint, all pollution would be reversed out of existence. Depending on the Refresher control settings, this effect could be world-wide.

"Raising up the foundations of many generations"

The Refresher would automatically rebuild previous decayed structures. It would rebuild and restore the entire earth.

Reversing forest fires

The Refresher not only would stop forest fires in its footprint, but would reverse the fires—restoring all that was lost—animate and inanimate, including lost trees and homes.

Reversing all calamities;
Reversing all effects of war;
Preventing all munitions from firing;
"Making wars to cease to the end of the earth."
Removing sinful propensities from people, including criminals;
Emptying prison houses;
Making possible and efficient Clean Energy Sources.
The blessings of the Refresher are endless. In short it would restore earth to Edenic perfection in under 3.5 years of machine time.

Chapter 6

Reversing the Sun

We could reverse the nova on the sun if we could reverse the second law of thermodynamics on the sun. It is not as difficult as it sounds. First we need to realize that the second law of thermodynamics on the sun is just the same as we are now experiencing it on earth. The order to disorder arrow of time in the second law of thermodynamics on the sun and earth now points from order to disorder. That is because the order energy on the sun and earth is now slightly negative. If we could make it slightly positive on the sun and earth, we could reverse the order to disorder arrow of time on the sun and earth, and therefore reverse the direction of processes on the sun and earth, on the sun making the Helium fission to Hydrogen again, backing up the Hydrogen to Helium processes on the sun. We could back up the sun several thousand years to safer times, before letting the sun go back to its Hydrogen to Helium fusion processes. We first need a basic understanding of the second law of thermodynamics and Order Energy E_O and Entropy Energy E_S.

Chapter 7

The Second Law of Thermodynamics

A. Introduction

Everything goes from a state of order to more disorder. Brand new automobiles wear out and rust. Objects break or are damaged. A thermos bottle falls off the counter, and the inner glass bottle is shattered. We do not expect the shattered bottle to fall back up to the counter and become whole again. There is a one-way arrow for the events to transpire. That arrow is the order to disorder arrow of time in the second law of thermodynamics.

Houses grow old and fall into decay. Barns fall down. Fruit spoils, people and animals grow old and die. Viruses mutate. People become ill and die. Crime and disorder in society increase. Homes break up. Aborted fetuses disintegrate. Dead people and things decompose. All of these negative occurrences are the outworking of the second law of thermodynamics—that part of which is an arrow making everything go from order to disorder.

Let us consider what other people have written about the second law of thermodynamics.

"*Second law of thermodynamics*
"An equilibrium macrostate of a system can be characterized by a quantity S (called *entropy*) which has the following properties:

"(i) In any infinitesimal quasi-static process in which the system absorbs heat dQ, its entropy changes by an amount

$$dS = \frac{dQ}{T} \qquad (7\text{-}1)$$

where T is a parameter characteristic of the macrostate of the system and is called its *absolute temperature*.

"(ii) In any process in which a thermally isolated system changes from one macrostate to another, its entropy tends to increase, i.e.,

$$\Delta S \geq 0. \qquad (7\text{-}2)$$

"The relation (7-1) is important because it allows one to determine entropy *differences* by measurements of absorbed heat and because it serves to characterize the absolute temperature T of a system. The relation (7-2) is significant because it specifies the direction in which non-equilibrium situations tend to proceed."[1]

The above expression of the second law of thermodynamics is regarding entropy and heat. Other writers include the order to disorder arrow in the second law of thermodynamics.

"It is a matter of common experience that disorder will tend to increase if things are left to themselves. (One has only to stop making repairs around the house to see that!) One can create order out of disorder (for example, one can paint the house), but that requires expenditure of effort or energy and so decreases the amount of ordered energy available.

"A precise statement of this idea is known as the second law of thermodynamics. It states that the entropy of an isolated system always increases, and that when two systems are joined together, the entropy of the combined system is greater than the sum of the entropies of the individual systems. For example, consider a system of gas molecules in a box. The higher the temperature of the gas, the faster the molecules move, and so the more frequently and harder they collide with the walls of the box and the greater the outward pressure they exert on the walls. Suppose that initially the molecules are all confined to the left-hand side of the box by a partition. If the partition is then removed, the molecules will tend to spread out and occupy both halves of the box. At some later time they could, by chance, all be in the right half or back in the left half, but it is overwhelmingly more probable that there will be roughly equal numbers in the two halves. Such a state is less ordered, or more disordered, than the original state in which all the molecules were in one half. One therefore says that the entropy of the gas has gone up. Similarly, suppose one starts with two boxes, one containing oxygen molecules and the other containing nitrogen molecules. If one joins the boxes together and removes the intervening wall, the oxygen and nitrogen molecules will start to mix. At a later time the most probable state would be a fairly uniform mixture of oxygen and nitrogen molecules throughout the two boxes. This state would be less ordered, and hence have more entropy, than the initial state of two separate boxes."[2]

"The explanation that is usually given as to why we don't see broken cups gathering themselves together off the floor and jumping back onto the table is that it is forbidden by the second law of thermodynamics. This says that in any closed system disorder, or entropy, always increases with time. In other words, it is a form of Murphy's Law: Things always tend to go wrong! An intact cup on the table is a state of high order, but a broken cup on the floor is a disordered state. One can go readily from the cup on the table in the past to the broken cup on the floor in the future, but not the other way round.

"The increase of disorder or entropy with time is one example of what is called an arrow of time, something that distinguishes the past from the future, giving a direction to time."[3]

B.   Electrino Model and 2nd Law

The natural tendency of leptons in beta decay is that the parent lepton combines with one or more gravitons to produce more particles. In all natural reactions, the order energy of the resultant particles is less than or equal to the order energy of the original particles.

1.   Negative Energies. Let us consider antimatter more carefully. "In the Dirac theory also, *the permissible energy values for a free particle range from $+mc^2$ to $+\infty$ and from $-mc^2$ to $-\infty$*. The first of these results is of course just what we expect for a free particle—that its total energy can have any value greater than its rest energy. But the second result is quite puzzling, since it implies the existence of states of *negative total energy*."[4]   Anderson in 1932 discovered positrons in cosmic radiation. These were regarded as Dirac's negative energy particles. "The first two solutions of the Dirac equation . . . clearly describe a free electron of energy E and momentum **p**. The two negative energy electron solutions . . . are to be associated with the antiparticle, the positron."[5]

However, in the annihilation it is not $(+mc^2) + (-mc^2) = 0$, but $2mc^2$ is the result of annihilation.[6] There is something strange going on with the minus signs in these equations. The calculations are inconsistent.

Maybe there are two kinds of energy considered. One we can call entropy energy E_S. In the annihilation reaction, $|+mc^2|+|-$

mc²| = 2mc². Entropy energy is the higher value. The other energy is order energy E_O. In order energy the same reaction is $(+mc^2) + (-mc^2) = 0$.

Let us consider entropy energy and order energy for particle decay schemes. There are a few decay schemes where no negative order energy (anti-matter) is introduced in the right hand side of the decay schemes. In those few instances, the final order energy is equal to the initial order energy (when kinetic energy is taken into account). But in most cases, a trace of negative order energy (anti-matter) is introduced into the right side of the decay schemes. There is nothing on the left hand sides of the decay schemes to correspond to this addition of a trace of negative order energy on the right sides of the decay schemes. Therefore, total order energy is less on the right hand sides of the decay schemes than on the left hand sides (if only by a trace). A few decay schemes introduce a lot of antimatter (as K⁻) on the right side of the decay scheme. The loss of order energy in the systems is greater in those cases. But in every case, for all natural processes, the order energy final is less than or equal to the order energy initial, or

$$\Delta E_0 \leq 0. \qquad (7\text{-}3)$$

Let us check the order energy for electron electrino fusion reactions. Electrons made energetic by acceleration (as heavy as protons) fuse and form anti-protons. Matter is converted to anti-matter. Entropy energy is conserved, but not so order energy. Order energy is reduced in the extreme from +938 MeV to -938 MeV or more for each electron fused (two electrons are fused in each reaction). The order-disorder arrow for electron electrino fusion points in the usual direction. The system does obey the second law of thermodynamics as we now know it.

2. Reversing the Order to Disorder Arrow. What would happen if we fused the electrino constituents of positrons instead of the electrino constituents of electrons? Entropy energy E_S would again be conserved. Entropy would be increased. However, order energy E_O would go from -2 x 938 MeV to +2 x 938 MeV—from disorder to order. The order to disorder arrow would be reversed. This would be a reaction that would be prohibited by the second law of thermodynamics—unless the strong gravitational force that fuses the anti-semions would be stronger than the second law of thermodynamics (which otherwise governs weak interactions), which it is.

Here we see that the entropy arrow of time and the order to disorder arrow of time are separate and distinct, and are not one and the same thing. While all the reactions the author has studied increase entropy, the fusion of positron anti-semions reverses the order to disorder arrow, making more order out of the disorder.

Positron constituent electrino fusion might not only take the electrinos from disorder to order. It could make other physical processes in a local area go from disorder to order. The positron fusion not only violates the second law of thermodynamics, it reverses the order to disorder arrow of that law in a local area, making other processes in that area reverse. Let us consider that process more to see how it might be regulated.

We guess the desired relationships for reversing the order to disorder arrow in the second law of thermodynamics through dimensional analysis. We want to solve for r, the maximum radius in which the reversed law would be effective. There is a way we can obtain a length from combinations of our variables and constants. That way is in the right hand side of Eq. (7-4). The whole expression is the thermodynamic relation we are seeking. The thermodynamic relation is:

$$(\Delta E_o)_t > 0 \text{ where } r < \frac{(\Delta E_o)_1 \, c}{ik}, \qquad (7\text{-}4)$$

where E_o is the order energy–the positive or negative energy in the pair production of particles; ΔE_o is the change in the order energy, where $(\Delta E_o)_t$ is the change in the total order energy of the system, and where **$(\Delta E_o)_1$** is the change in the order energy for a single source reaction—for a positron fusion reaction it is approximately 2 x 0.94 x 10^9 eV/collision x 1.6 x 10^{-19} joules/eV = **3.0 x 10^{-10} joules/collision**; c is the speed of light—approximately **3.0 x 10^8 m/s**; we shall solve for the effective radius r; **i** is the effective beam collision current in each beam in Coulombs per second (we will solve for 10^{-11} or 10 picoAmps); **k** is the ratio of particle energy to particle charge. This energy per charge is the accelerated energy of the particle (0.94 x 10^9 ev times 1.6 x 10^{-19} joules/ev = 1.5 x 10^{-10} joules) divided by the charge of each positron (q = 1.6 x 10^{-19} coulombs), which equals **9.38 x 10^8 joules per coulomb**. The collision efficiency eff is not needed in this equation, because the result is not in particles, but is already in collisions.

Incredibly, the lower the collision rate, the bigger the radius of the affected area. And the greater the collision rate, the smaller the radius of the affected area. With 10^{-11} A effective beam currents (at 100% efficiency), the effective radius **r** solves for **9.6 meters**—which describes a small area—less than a tenth of an acre. Instead of beam current collisions at 100% efficiency, we calculate the net collision rate needed, which could be achievable with much lower particle collision efficiency.

To get an idea of the positron collisions needed to reverse the order to disorder arrow of the second law of thermodynamics in what size of affected radius, see Table 7-1 below.

For footprint the radius of	r	effective collision current	net collisions/sec
house	9.6 m	10 pA	6.2E7
4 ftball flds	96 m	1 pA	6.2E6
community	960 m	100 fA	6.2E5
city	9.6 km	10 fA	6.2E4
Israel	160 km	0.6 fA	3,800
U.S.	2,400 km	0.04 fA	250
World	13,000 km	0.008 fA	46
Sun	1.7E11 m	6E-22 A	1 collision/79 hrs

Table 7-1. Net collision rate and effective collision current of positrons at 940 MeV colliding with spin flipped positrons at 940 MeV necessary to make a Refresher footprint of given radius for reversal of the order to disorder arrow of the second law of thermodynamics. [To calculate the net collisions/sec, multiply the net equivalent collision currents by 6.25E18.]

Remarkably enough, the affected area of second law reversal calculates to increase with the reduction of positron effective beam current. Area control is merely a matter of timed gating of the positrons in the positron-positron collider.

---

[1]F. Reif, *Statistical Physics*, Berkeley Physics Course—Volume 5 (New York: McGraw-Hill Book Company, 1967), p. 283.
[2]Stephen Hawking, *A Brief History of Time*—From the Big Bang to Black Holes (New York: Bantam Books, 1988), pp. 102, 103.

[3]*Ibid.*, pp. 144, 145.

[4]Robert B. Leighton, *Principles of Modern Physics* (New York: McGraw-Hill Book Company, Inc, 1959), p. 665.

[5]Francis Halzen, Alan D. Martin, *Quarks and Leptons* (New York: John Wiley & Sons, 1984), p. 107.

[6]David S. Saxon, *Elementary Quantum Mechanics* (San Francisco: Holden-Day, 1968), p. 386.

Problem Set

1. "Humpty-Dumpty sat on a wall. Humpty-Dumpty had a great fall. And all the king's horses and all the king's men couldn't put Humpty-Dumpty back together again." What law of physics does this child's nursery rhyme illustrate? What arrow of time is demonstrated?

2. Is the entropy more or less after two gas containers are opened to each other?

3. You watch a movie. Broken pieces of glass fly up and become a window. Is the movie playing forward or backward? How do you know?

4. What generally happens in beta decay? Is the system going to more or less order?

5. What would the world be like if the order to disorder arrow were reversed?

Chapter 8

The Theory

Refresher

    The Principal Investigator has discovered a new Grand Unification Theory (GUT). It has deeper symmetry and lower orbital structures than the Standard Model of Physics. It has greater parsimony than the Standard Model. Whereas the Standard Model requires 61 different elementary particles to construct known light and matter [1](page 48), the GUT requires only two different elementary particles to construct known light, matter, and gravitons; and those two different particles can both be ionized from empty space with a single particle, and can combine in a single particle.

    What differences does this GUT have to the Standard Model? The GUT is an aether model of physics. It has aether special and general relativity, rather than Einstein's aether-less Special and General Relativity. This makes a simple model of gravity and inertia possible [2](Chapter 5). Up until now, uniting special and general relativity in particle physics has been as difficult as uniting fire and ice. This problem is solved with aether special and general relativity in the GUT [2](Chapter 6). Special and general relativistic calculations are both exact fits in the particle structures calculated in that chapter.

    The GUT has one postulate that states that symmetric smooth charge distributions cannot have detectable spin. But electrons and positrons have detectable spins. Therefore they must not be symmetric point charges, but have two half charges in them orbiting about each other. The orbiting like charges show that fracton charges come in $\pm e$, $\pm e/2$, $\pm e/4$, and $\pm e/8$ (the Electrino

Hypothesis), rather than in ± e/3 and ± 2e/3 (the Quark Hypothesis). The Electrino Hypothesis is very different from the historic and accepted Quark Hypothesis. Yet it does not lead to untenable particle structures. The Principal Investigator has induced the particle structures of all known light, matter, and gravitons, through the simultaneous satisfaction of ten criteria: particle charge, spin, parity, mass, spin feasibility, preceding particles (to avoid duplication), the Pauli Exclusion Principle, b state laws, decay schemes, and the requirement that no particles except electrons and positrons and neutrinos and anti-neutrinos have ground state − and + echons in them, so no particles other than electrons and positrons have to have defined masses in them. [2](Appendix B). Satisfying all the listed Decay Modes published in *Summary Tables of Particle Properties*, by the Particle Data Group [3] (which reference [2](Appendix B) does), is the satisfaction of thousands of tests. Except for being an unknown model, the GUT is in a strong position. Its particle structures are all unique. The quark model particle structures are not all unique. The quark model has seven formulas for eight particles. The first and the last are the same. But the masses of the first particle and the last particle are not the same! The quark model fails.

Another difference of the GUT to the Standard Model is that charged sub-particles of like charge fractions can fuse to particles of higher charge fractions. [2](Chapter 12). The secret of why that should be is that when sub-particles orbit or travel faster than the speed of light in the relativistic frame relative to the baseline non-relativistic frame, their radii become imaginary because of the relativistic length contraction formula. The strong electric force equation for these super luminal sub-particles has two such imaginary radii multiplied together. That makes an additional minus sign in the force equation, which makes like charges attract. When two bound sub-particles of a positron collide in the Center of Mass Frame with two other bound sub-particles of another positron with like oriented spins in the Center of Mass Frame, with 1880 MeV energy or more, the four sub-particles are attracted into the same orbit. Then one sub-particle from one positron is more attracted to one sub-particle from the other positron than to any other sub-particle because of closer proximity. These are like charges, and here like charges attract because they travel faster than light. [Einstein predicted that nothing could go faster than the speed of light. But neutrinos have recently been clocked as going faster than the speed of light. [4-6] Some Einstein loyalists have attributed those results to various experimental errors, but it is impossible to have a Unified Field Theory without some particles traveling faster than the speed of light.] The two sub-particles are attracted by the electric strong force. Nothing stops them from fusing. The other two positron sub-particles fuse also. Four sub-particles fuse down to two particles each with twice the charge as the charge of one of the four sub-particles. The four sub-particles are ½ e charges. The two fused particles are 1 e charges each, but though they are numerically whole particles, they cannot exist alone. That is why on creation they scavenge from the graviton sea the necessary sub-particles to become protons or neutrons. But when the positron four ½ e sub-particles fuse to the two 1 e particles, they switch from antimatter to matter. The fusion of sub-particles in positrons results in the generation of solely positive order energy (quantum mechanical energy in the creation of particles). This phenomenon is theorized to reverse the order to disorder arrow in the second law of thermodynamics [because it is positive order energy as opposed to negative order energy which surrounds us and which determines the current order to disorder arrow direction and the direction of reactions]. [2](Chapter 16).

The fusion of the sub-particles of positrons can result in the reversal of the order to disorder arrow in the second law of thermodynamics—but over what distance? Those answers are already derived in *Refresher* 1, Draft 2, Chapter 2 (pages 13-15).

Incredibly, the lower the effective current (or the lower the collisions per second) the bigger the radius of the affected area. And the greater the effective current, the smaller the radius of the affected area. With 10^{-11} Amps effective collision currents, the effected radius r solves for 9.6 meters, which describes a small area—less than a tenth of an acre.

To get an idea of the positron net collision currents needed to reverse the order to disorder arrow of the second law of thermodynamics in what size of affected radius, see Table 7-1 in the previous chapter.

We don't need this in the Clean Energy Source, but the Refresher needs electronics to gate the colliding beams by at least thirteen decades with a control board key-lock position for each of switch settings decades below 100 pA/eff net collision rates. (To obtain the collisions per second, multiply the effective collision currents by 6.25E18.) Initially we will not know the efficiency eff. We will not be correct, but until we can measure the efficiency of collisions, we can assume it is 1.0. You can use as much beam power as you want up to 1.0 Amp to achieve initial collisions, but then gate it down to 100 pA; 10 pA 1.0 pA; 0.1 pA; 10 fA; 1.0 fA; 0.1 fA; 0.01 fA; 0.001 fA; 1.0E-19 A; 1.0E-20 A; 1.0E-21 A; and 1.0E-22 A. The gating needs to be before the beams collide. The spin flipper on one of the beams appears to be the best place to gate the beams: If you don't flip the beam for a time, or you flip it less than 90 degrees, the beam will shoot blanks, or essentially the beam will be chopped for a time. For reference sake let beam current height times total beam current width be 1.0 pA/eff for beam collisions for on the order of 100 meters radius foot print.

Let lower net beam currents be same current heights and peak widths, but with longer times between gated peaks. This will increase footprint radius. (It is the inverse of what you would think. It is counter intuitive.) It is essentially an electronic clock hooked to the beam flipper.

The author will now calculate the rate at which reverse aging will occur in the calculable radius of the active Refresher: The beginning energy of the particles (positrons) from which the fusion process takes place is $2m_e c^2$ per individual reaction. The ending energy of the particles (protons) to which the fusion process tends is $2m_p c^2$ per individual reaction. $\dfrac{\Delta E_p}{\Delta E_{e^+}} = \dfrac{+2m_p c^2}{-2m_e c^2} \approx -1836$. This is a unit less expression from the available energy terms. What we seek is another unit less expression $\dfrac{\Delta t_r}{\Delta t}$, where t is the normal time during which a person or object ages, and t_r is the reverse time (negative) during which a person or object un-ages. The quotient is the relative rate of un-aging compared to aging. This also is a unit less quotient. What use of particle fusion parameters can yield such a unit less quotient? What terms are available to derive such a unit less quotient? What about the first terms and unit less quotient? If we equate them, we have $\dfrac{\Delta t_r}{\Delta t} \approx -1836$. Reverse time would be negative and 1836 times as fast as forward aging time. Forward aging of 100 years would be un-aged in 19.89 days. Forward aging of 1 year would be un-aged in approximately 4.77 hours of machine time.

The Refresher would have many different effects in a controlled area, and the controlled area could be varied in size from the thickness of pencil lead or less to the earth orbit of the sun. Some such effects would be reverse aging; backing diseases out of existence; backing decay and pollution out of existence; disaster and war relief; making a demilitarized zone in the controlled area where no explosives will explode; building and vehicle maintenance; making a new form of food preservation; removing criminal tendencies in brains; taming animals; and making Clean Energy Sources possible; etc.

Placing a Clean Energy Source in the footprint of the Refresher would prevent the creation of radioactive wastes in the Clean Energy Source, because the force of the positive sense disorder to order arrow in the Refresher would be about 1836 times as powerful as the negative sense order to disorder arrow in the Clean Energy Source. The power of the Refresher disorder to order arrow would over power the force of the Clean Energy Source order to disorder arrow.

But would the power of the Refresher disorder to order arrow of time prevent the sub-particles of electrons in the Clean Energy Source from fusing? The answer is no! The gravitational and electrical strong forces attracting the semions in electrons in the Clean Energy Source are the strongest forces of all—stronger than the positive sense disorder to order arrow in the Refresher! Where we want radioactive wastes to be suppressed in the Clean Energy Source, they are suppressed. And where we want the fusion processes not to be suppressed, they are not suppressed! The Refresher and the Clean Energy Source appear to be made for each other!

    1. How can scattered glass fragments be re-gathered together and fused into a perfect window without the actual backing up of the clock? The order to disorder arrow in the second law of thermodynamics is a powerful force: it is more powerful than gravity; it is more powerful than the force of a speeding projectile; it

is more powerful than chemical bonds; it is more powerful than surface tension or friction; it is more powerful than atomic or hydrogen bombs! It is so powerful that nothing that science could do could resist it—up until now! For thousands of years the whole human race has been controlled and defeated by the order to disorder arrow of time turned in the negative sense. But there is something in nature that is more powerful than the order to disorder arrow in the second law of thermodynamics! It is the strong electric and strong gravitational forces. But their action is confined in extremely small spheres. How could they be controlled and made to work in man's benefit? The aether is the key here! Einstein's aether-less relativities will not work here! The fusion of electrinos—anti-semions into unitons in those extremely small spheres is accomplished by those resistless and non-reversible strong forces. The process produces positive order energy and the creation of new protons. The aether penetrating those extremely small spheres pass through and somehow carry away the information that new protons are being created in this vicinity. That information, carried by the outgoing aether, flips the order to disorder arrow to the positive sense, which now is a super powerful force to re-create everything! An array of broken glass fragments has an order to disorder value. In the active area of the Refresher or Regenerator machine [for which we have already calculated its effective radius based on effective collision currents], the system is forced in a small increment of time to move the glass shards to a slightly more positive order to disorder value. It keeps doing this one Δt and one ΔE_O at a time until the glass shards fall back up and fuse into the original window. This type of process will work for re-uniting severed spinal cords and healing and restoring war wounds.

2. Cystic fibrosis and Down's syndrome victims are cured by another process of this order to disorder reversal. The DNA code that God made for every living thing is a powerful self-correcting thing from most mutations. But not in the cases of Cystic fibroses and Down's syndrome victims in the current negative order to disorder arrow sense. But when the order to disorder arrow is flipped to the positive sense by the creation of new protons, there is a powerful force in all the bodies to restore the DNA and RNA codes in every cell to the maximum order correct codes. Every mutation of the DNA and RNA codes is deleterious, as was demonstrated by the irradiation of fish egg experiments in the Fisheries Department of the University of Washington USA in the 1970s. Irradiation of fish eggs produced thousands of mutations and deformities—but not one beneficial mutation. There are gillions of wrong code values, but only one maximum order code for every conception code. When the order to disorder arrow in the second law of thermodynamics is flipped to the positive sense, there is a powerful force in every cell of the body to right the existing codes to the correct codes. That is why the DNA and RNA codes in every cell are righted to the original code in each cell to the same code. This heals Cystic fibrosis and Down's syndrome cases as well as not desired skin colors and

birth deformities. Everyone's codes are restored to the correct conception codes.

3. Every disease germ or cell or virus has negative code values. The good pro-biotic bacteria all have positive codes. Reversing the order to disorder arrow in people and their environs backs the harmful bacteria and viruses out of existence without harming the good bacteria as antibiotics do. This can back almost every disease out of existence including HIV AIDs, cancer, Ebola and Zika viruses.

But how can we penetrate those extremely small spheres confining the strong forces, to induce an anti-semion to uniton fusion reaction? Normally electron or positron beams have a 1.602E-19 efficiency of colliding. But microscopic fine beams even at that efficiency can have one or two or a few collisions. This collision efficiency works for electrons as well as positrons; but for electrons it results in a negative sense for the order to disorder arrow, which can be reversed only by a stronger positive sense of the order to disorder arrow from the fusion of positron anti-semions to unitons, the core particles of protons and neutrons. But fortunately the positive sense of the order to disorder arrow in the second law of thermodynamics from the fusion of positron anti-semions to unitons is about 1836 times as powerful as the negative sense of the order to disorder arrow of time in the second law of thermodynamics by the fusion of electron semions to anti-unitons. This permits Clean Energy Sources to work in the presence of a Refresher. But back to the penetration of the extremely small spheres by positrons: it just takes one collision above 1876 MeV in the center of mass frame [with axial spins, with the spins from one positron accelerator flipped relative to that accelerator, so that the semions from both accelerators have the same sense on colliding] to send the aether outward in a diverging rays road map focusing all successive positrons to that location of proton creation. The diverging aether rays make a three dimension and time roadmap not only in the outer non-relativistic frame, but also in the inner relativistic imaginary dimensions into the extremely small [on the order of E-43 imaginary meters] spheres containing the strong forces. Just one anti-semion fusion event [four semions to two unitons] creates a special roadmap for successive positron anti-semion replications. But remember that just one fusion event will affect a much larger volume in reversing the order arrow, than a few or many fusion events. The efficiency of the events is just the inverse of what we expect. The system is counter intuitive. On starting the chain of events, there would almost certainly be a solar system wide positive order energy transient, but it would not necessarily be long lasting.

When we stop to ponder the restorative processes, they are less like unbelievable miracles, and more like desirable understood processes. Reversing forest fires would be the first type of process on a grand scale. Conceptually it is simple rapid Δt, ΔE_O movements of particles and gases, recombining them according to their original codes.

Do these added insights into electrino fusion restorative processes help any in relieving disbelief in the seemingly miraculous impossibilities?

The Refresher would be quite versatile: It could be built on land, in or on ships, in large airplanes, in long rail cars, in long semis, and in space craft and space bases. It would revolutionize the earth and would bring peace to earth.

[1] David Griffiths, *Introduction to Elementary Particles (New York: John Wiley &Sons, Inc., 1987).*

[2] Gordon L. Ziegler, *Electrino Physics*, Draft 2 (obtainable from amazon.com).

[3] C. Caso, *et. al.* (Particle Data Group), "Summary Tables of Particle Properties," (including "Gauge and Higgs Bosons Summary Table," "Lepton Summary Table," "Meson Summary Table," and "Baryon Summary Table), *CRC Handbook of Chemistry and Physics*, 80th Edition, David R. Lide, Ph.D., Editor-in-Chief (Boca Raton: CRC Press, 1999-2000), pp. **11**-1 to **11**-42.

[4] Frank Jordans and Seth Borenstein, "Neutrinos clocked moving at faster-than-light speed" (Associated Press): http://www.msnbc.msn.com/id/44629271/ns/technology_and_science-science/?gt1=43001

[5] Robert Evans, "Particles found to break speed of light" (Reuters): http://www.reuters.com/article/2011/09/22/us-science-light-idUSTRE78L4FH20110922

[6] "Do neutrinos move faster than the speed of light?" – physicsworld.com: http://physicsworld.com/cws/article/news/47283

Chapter 9

Is It Safe?

Current science has no data on this, but religious sources have rather a lot to say on this subject, in reassuring the safeness about reversing the order to disorder arrow in the Second Law of Thermodynamics. (See Gordon L. Ziegler, *Regenerating the Sun and Earth Through God Inspired Science*, Chapter 4; and Gordon L. Ziegler, *Does God Really Love Us?*, Chapter 8.)

Chapter 10

Should We Do It?

Should we build and operate an order to disorder arrow in the second law of thermodynamics reverser (Refresher)? Apparently to do so would be safe. But why should we do it?

People are spending thousands upon thousands of dollars for medical diagnostics and treatments that do not cure, where they could have a complete cure for free if the Refresher were operating. The suffering and dying of people weigh upon the author—especially the suffering and dying of people interested in the building and operating of the Refresher.

But people have suffered and died for thousands of years. Why should that change now? Why not thousands of years ago? It is only recently that science has been advanced enough to build and operate the Refresher. And besides, science has taken some wrong turns, and is entrenched in wrong theories. This breakthrough would not be possible without the discovery of a new Grand Unification Theory of Physics, which has occurred to the author in recent years.

Without the creation and operation of a Refresher quickly, the sun will peak seven times brighter and hotter soon, killing millions of people. Shouldn't we fund and build a Refresher quickly? Even if we survived the nova peak without a Refresher, the sun would soon go out in darkness. This would be far worse than the nova peak! If we wanted to survive this, we should not only make one or more Refreshers, but mass produce Clean Energy Sources on earth, making the light and heat of the sun unnecessary. But what we do, we need to do quickly!

Chapter 11

Clean Energy Sources

Electrino fusion can not only supply perfect health, soundness of mind and body, and eternal life, it can supply the human race and the Universe with free, absolutely clean energy— no Carbon emissions, no radioactive wastes, no wastes at all, and little or no heat pollution! It is 1000 times as efficient as nuclear power; it can go 100 or 200 years without refueling; it doesn't require dangerous radioactive fuels or dangerous chain reactions. It can safely be shut off at a flip of a switch with no decay heat. It has no hazard of a meltdown or a radiation accident. It is very efficient. The annihilation of a single penny would produce $1,872,000 worth of electricity at bargain electricity rates of $0.03/kwh. The first model would use inexpensive common brass for fuel. Whereas a new nuclear power plant would cost $6.0 billion and have 1250 MW AC output, rf cavity particle acceleration Electrino Fusion Power (EFP) Power would cost $110 million for the prototype and $50 million or less for successive plants and would produce 1880 MW maximum (or less if desired) each plant DC which could be converted to AC and put on existing electricity grids. Whereas a new nuclear power plant would take 10 years to build, the rf cavity style EFP power plant would take only one year to build or less and test, and could be pre-fabricated and proliferated around the world quickly.

There is now an alternate less expensive and faster way to build an EFP Reactor—through laser particle acceleration. It may cost only $5 million, and would take only four months to design and build.

Apart from the construction costs, EFP energy could operate for free with a few thank offerings. Were the construction financed by loans, the loans could be paid fourfold with $0.03/kwh in one year of operation, then free electricity afterwards.

To replace the solar energy on earth, to save the earth if the sun goes out at the end of a nova, we would have to produce 4000 times as much energy as the total earth electric production currently. And this would have to be done quickly. We would have to manufacture mass produce the Clean Energy Sources and proliferate them around the earth quickly, with training conventions of engineers, financiers, laymen and children in the great centers of population around the earth.

Chapter 12

Clean Energy Theory

1. Introduction

The technical name for the Radioactive Waste-free Reactor is the Electrino Fusion Power Reactor (EFP Reactor). Electrino is the author's name for tiny electric particles that compose all light, matter, and gravitons in the author's new Grand Unification Theory (GUT). The main difference between the Standard Model and the new GUT is that fracton charges in the GUT come in ± e, ± e/2, ± e/4, and ± e/8; whereas fracton charges in the Standard Model come in ± 2e/3 and ± e/3. The change in fracton charges did not lead to untenable particle structures. The author induced the structures of every known particle according to the scheme in the GUT. They all worked out all right. And whereas it takes 61 elementary particles to build known light and matter in the Standard Model, it takes only one according to the GUT. The GUT has deeper levels of symmetry and lower orbits. This chapter develops the features of the radioactive waste-free EFP Reactor using the new GUT.

2. Elementary Particle Fusion

In the new GUT (which, by the way, is called Electrino Fusion Model of Elementary Particles), the particles are held together by symmetrical orbits in elementary particles (not in neuclei), not glued together by gluons. The quarks, with ± 2e/3 and ± e/3 fracton charges, do not lend themselves to stable, symmetrical orbits, but the electrinos, with ± e, ± e/2, ± e/4, and ± e/8 fracton charges, do. In the model, photons are composed of heavy positive and negative whole charges orbiting about each other, and traveling together at the speed of light; electrons are made up of like light half charges orbiting about each other; and pions are made up of two orbiting pairs of like light fourth charges orbiting in the opposite directions, superimposed on each other. Notice the symmetry. Notice the orbits. Notice the space between the particles. Notice the individuality of the particles—bound only by the speed of light barrier and orbital mechanics.

It is important to notice the velocities of the particles and their behaviors at those velocities. All fractons (called electrinos in the model) travel either just slightly faster than the speed of light, or significantly faster than the speed of light. The point is, they all travel faster than the speed of light. For the light ones, this affects their radii—making them imaginary. This affects their force. Whereas slow like-charges repel, faster than c like-charges attract. This affects the potential energy of particles. This makes deep potential wells at the top of potential hills for the potential energy of charged particles. This affects the perceived mass-energy of the particles—positive instead of negative.

Faster than c like-charges attract. Negatively charged like half charges traveling just faster than c orbit around each other forming electrons. If the electrons never collide with any other electrons—at least not with sufficient energies—the half particle inertias in them cause the half charges to orbit always opposite each other—never approaching each other. But if electrons collide with each other with over 938 MeV each in the same orbit, four half charges come near to each other. The four half charges are not all held opposite each other. They all attract each other. What will happen? One half charge from one electron will be attracted to one half charge from the other electron. Nothing will stop the half charges. They will travel until they contact each other. What happens then? They are like charged. They form a new particle with twice the half charge—in other words a whole charge. We could say the half charges fuse to a whole charge.

When high energy electrons collide, not only do two half charges from opposite electrons fuse, the other two half charges on the opposite side fuse. We have four half charges from two electrons fusing to two whole charges. What then?

It is profitable at this juncture to assign fracton or electrino structures to simple particles. Pions are composed of four positive fourth charges—two orbiting one way, the other two orbiting the opposite way, superimposed on each other. Electrons are made of two light weight negative half charges. Neutrons are constructed of a heavy positive whole particle orbited by an electron. If the constituents of pions were fused to the constituents of electrons, it would be to positive electrons—positrons—antimatter. If the sub-particles of negative electrons were fused to the heavy whole core particles of neutrons, it would be to negative neutrons—antimatter. If we started with the opposite charges of above, the particles would fuse to matter instead of antimatter. Every time there is a fusion of electrinos, there is a switch from matter to antimatter or vice versa.

What would happen to the negative half charges in electrons fused to whole particles above? The half charges would be negatively charged matter. The whole charges would be negatively charged core particles of antimatter—anti-protons and anti-neutrons. The anti-core-particles would scavenge from the graviton sea the remaining portions of anti-protons and anti-neutrons. The resultant anti-protons and anti-neutrons would drift into local protons and neutrons and annihilate them, giving off gamma rays, which could be converted into electricity. This is the foundation of the science of the radioactive waste-free EFP Reactor. The electricity comes from processed gamma rays, which come from the annihilation of protons and anti-protons and neutrons and anti-neutrons, which come from anti-protons and anti-neutrons, which come from negative heavy whole core particles (antimatter), which come from the fusion of half particles in electrons, which come from the collision electrons above 938 MeV each electron, with like spins in the center of mass frame.

3. Efficiencies

Before electrons can have fusion of their half particles, they must be accelerated to at least the masses of protons—938.27231 ± 0.00028 MeV [1]—roughly at least 939 MeV. That is a necessary energy investment into the process. When the particles fuse, there follows an annihilation of both a proton and an anti-proton or a neutron and an anti-neutron. Nearly twice as much energy in gamma rays results as was invested in the acceleration of electrons. At first this sounds good. But then we realize we must be more than 50 per cent efficient over-all in order to be self-sustaining and be an energy source using this energy phenomenon. That is hard to achieve. State of the art accelerator efficiency in 1988 was itself only 50% [2]. While individual steam turbine efficiencies were as high as 96.1%, the world record steam turbine gross efficiency recently was 48.5% [3]. That is an overall efficiency for our process of less than 24.25%. And we need 50% to break even, let alone have a surplus to become a new power source!

4. A Surprising Turn

The lack of necessary efficiency of the fusion-annihilation reaction is discouraging. The author put this process on the back burner until he would receive greater light upon the subject. Things took a surprising turn. Through fusing the sub-particles of positive electrons—positrons—in theory, he learned how to reverse the order to disorder arrow in the second law of thermodynamics. That is huge! That is a way to reverse aging, disease, and decay processes—to make old people young again and back out all diseases from existence! Let us read what he first wrote about the process and the phenomenon.

"The explanation that is usually given as to why we don't see broken cups gathering themselves together off the floor and jumping back onto the table is that it is forbidden by the second law of thermodynamics. This says that in any closed system disorder, or entropy, always increases with time. In other words, it is a form of Murphy's law: Things always tend to go wrong! An intact cup on the table is a state of high order, but a broken cup on the floor is a disordered state. One can go readily from the cup on the table in the past to the broken cup on the floor in the future, but not the other way round.

"The increase of disorder or entropy with time is one example of what is called an arrow of time, something that distinguishes the past from the future, giving a direction to time." [4]

5. Electrino Model and 2nd Law

The natural tendency of leptons in beta decay is that the parent lepton combines with one or more gravitons to produce more particles. In all natural reactions, the order energy of the resultant particles is less than or equal to the order energy of the original particles.

a. Negative Energies. Let us consider antimatter more carefully. "In the Dirac theory also, *the permissible energy values for a free particle range from $+mc^2$ to $+\infty$ and from $-mc^2$ to $-\infty$*. The first of these results is of course just what we expect for a free particle—that its total energy can have any value greater than its rest energy. But the second result is quite puzzling, since it implies the existence of states of *negative total energy*." [5] Anderson in 1932 discovered positrons in cosmic radiation. These were regarded as Dirac's negative energy particles. "The first two solutions of the Dirac equation . . . . clearly describe a free electron of energy E and momentum **p**. The two negative energy electron solutions . . . are to be associated with the antiparticle, the positron." [6]

However, in the annihilation it is not $(+mc^2) + (-mc^2) = 0$, but $2mc^2$ is the result of annihilation. [7] There is something strange going on with the minus signs in these equations. The calculations are inconsistent.

Maybe there are two kinds of energy considered. One we can call entropy energy E_S. In the annihilation reaction, $|+mc^2| + |-mc^2| = 2mc^2$. Entropy energy is the higher value. The other energy is order energy E_O. In order energy the same reaction is $(+mc^2) + (-mc^2) = 0$.

Let us consider entropy energy and order energy for particle decay schemes. There are a few decay schemes where no negative order energy (anti-matter) is introduced in the right hand side of the decay schemes. In those few instances, the final order energy is equal to the initial order energy (when kinetic energy is taken into account). But in most cases, a trace of negative order energy (anti-matter) is introduced into the right side of the decay schemes. There is nothing on the left hand sides of the decay schemes to correspond to this addition of a trace of negative order energy on the right sides of the decay schemes. Therefore, total order energy is less on the right hand sides of the decay schemes than on the left hand sides (if only by a trace). A few decay schemes introduce a lot of antimatter (as K⁻) on the right side of the decay scheme. The loss of order energy in the systems is greater in those cases. But in every case, for all natural processes, the order energy final is < the order energy initial, or

$$\Delta E_0 \leq 0. \qquad (12\text{-}1)$$

Let us check the order energy for electron electrino fusion reactions. Electrons made energetic by acceleration (as heavy as protons) fuse and form anti-protons. Matter is converted to anti-matter. Entropy energy is conserved, but not so order energy. Order energy is reduced in the extreme from +938 MeV to -938 MeV or more for each electron fused (two electrons are fused in each reaction). The order-disorder arrow for electron electrino fusion points in the usual direction. The system does obey the second law of thermodynamics as we now know it.

2. Reversing the Order to Disorder Arrow. What would happen if we fused the electrino constituents of positrons instead of the electrino constituents of electrons? Entropy energy E_S would again be conserved. Entropy would be increased. However, order energy E_O would go from -2 x 938 MeV to +2 x 938 MeV—from

disorder to order. The order to disorder arrow would be reversed. This would be a reaction that would be prohibited by the second law of thermodynamics—unless the strong gravitational force that fuses the anti-semions would be stronger than the second law of thermodynamics (which otherwise governs weak interactions), which it is.

Here we see that the entropy arrow of time and the order to disorder arrow of time are separate and distinct, and are not one and the same thing. While all the reactions the author has studied increase entropy, the fusion of positron anti-semions reverse the order to disorder arrow, making more order out of the disorder.

Positron constituent electrino fusion might not only take the electrinos from disorder to order. It could make other physical processes in a local area go from disorder to order. The positron fusion not only violates the second law of thermodynamics, it reverses the order to disorder arrow of that law in a local area, making other processes in that area reverse. Let us consider that process more to see how it might be regulated.

We guess the desired relationships for reversing the order to disorder arrow in the second law of thermodynamics through dimensional analysis. We want to solve for r, the maximum radius in which the reversed law would be effective. There is a way we can obtain a length from combinations of our variables and constants. That way is in the right hand side of Eq. (12-2). The whole expression is the thermodynamic relation we are seeking. The thermodynamic relation is:

$$(\Delta E_o)_t > 0 \text{ where } r < \frac{(\Delta E_o)_1 \, c}{ik}, \qquad (12\text{-}2)$$

where E_o is the order energy–the positive or negative energy in the pair production of particles; ΔE_o is the change in the order energy, where $(\Delta E_o)_t$ is the change in the total order energy of the system, and where $(\Delta E_o)_1$ is the change in the order energy for a single source reaction—for a positron fusion reaction it is approximately 2 x 0.94 x 10^9 eV/collision x 1.6 x 10^{-19} joules/eV = **3.0 x 10^{-10} joules/collision**; **c** is the speed of light—approximately **3.0 x 10^8 m/s**; we shall solve for the effective radius r; **i** is the effective beam collision current in each beam in Coulombs per second (we will solve for **10^{-11} A** or 10 picoAmps); **k** is the ratio of particle energy to particle charge. This energy per charge is the accelerated energy of the particle (0.94 x 10^9 ev times 1.6 x 10^{-19} joules/ev = 1.5 x 10^{-10} joules) divided by the charge of each positron (q = 1.6 x 10^{-19} coulombs), which equals **9.38 x 10^8 joules per coulomb**. The collision efficiency eff is not needed in this equation, because the result is not in particles, but is already in collisions.

Incredibly, the lower the current, the bigger the radius of the affected area. And the greater the current, the smaller the radius of the effected area. With 10^{-11} A equivalent collision beam currents, the effected radius r solves for 9.6 meters.

To get an idea of the positron collisions needed to reverse the order to disorder arrow of the second law of thermodynamics in what size of affected radius, see Table 7-1.

Remarkably enough, the affected area of second law reversal calculates to increase with the reduction of positron beam current. Area control is merely a matter of timed gating of the positrons in the positron-positron collider. [8]

6. Rate of Reversed Aging

The author will now calculate the rate at which reverse aging will occur in the calculable radius of the active Refresher: The beginning energy of the host particles (positrons) from which the fusion process takes place is $2m_ec^2$ per individual reaction. The ending energy of the host particles (protons) to which the fusion process tends is $2m_pc^2$ per individual reaction. $\frac{\Delta E_p}{\Delta E_{e^+}} = \frac{+2m_pc^2}{-2m_ec^2} \approx -1836$. This is a unit less expression from the available energy terms. What we seek is another unit less expression $\frac{\Delta t_r}{\Delta t}$, where t is the normal time during which a person or object ages, and t_r is the reverse time (negative) during which a person or object un-ages. The quotient is the relative rate of un-aging compared to aging. This also is a unit less quotient. What use of particle fusion parameters can yield such a unit less quotient? What terms are available to derive such a unit less quotient? What about the first terms and unit less quotient? If we equate them, we have $\frac{\Delta t_r}{\Delta t} \approx -1836$. Reverse time would be negative and 1836 times as fast as forward aging time. Forward aging of 100 years would be un-aged in 19.89 days. Forward aging of 1 year would be un-aged in approximately 4.77 hours of machine time.

7. Miracle Working Power of the Refresher 1

The theoretical discovery of the order to disorder arrow in the second law of thermodynamics reverser (Refresher 1 for short) was a surprising turn, and engrossed the author for several years. By simply reversing the natural arrows between ordered events, many miraculous results were found to take place in theory.

What does it mean that the order to disorder arrow in the second law of thermodynamics is reversed? Events naturally come in order indicated by the arrows:

Healthy young adult→aging→wrinkles→aging→cancer→death→cremation→scattering ashes

Reversing the order to disorder arrow in the second law of thermodynamics means all the arrows between the ordered events are turned around. The old and diseased become young and healthy. The clock is not really reversed. Adults do not become children again and disappear to extinction. The system just tends to maximum order, which is at young adulthood. Children still grow up to maximum order at young adulthood.

Many similar reversals can occur in the animal kingdom and the environment. The author imagined many marvelous things, but virtually forgot about the EFP Reactor.

8. EFP Reactor in the Field of the Refresher 1

Finally the thought came, "What would occur if the EFP Reactor were in the field of a Refresher? The concepts of the effects assembled slowly. The accelerator electronics would not have resistive heating in the field. As a result the accelerator would be room temperature superconductive. There would not be any need for cryogenic energy losses. The accelerator would be about 100% efficient.

Reversing the order to disorder arrow in the second law of thermodynamics greatly affects all things with which we are familiar. But what would it do to photovoltaic cells in a high energy gamma field? Outside the Refresher field, photovoltaic cells in the high energy gamma field would become damaged. They would become more and more damaged with time. This is a form of aging. What would happen if the aged photovoltaic cells were put in an order reversed Refresher field? The cells would un-age back to the original condition. What would happen if photovoltaic cells in an order reversing Refresher field were exposed to high level gamma radiation? They would not become damaged or aged. What would happen to the power that would ordinarily be absorbed in the aging process? Would it not be added to the power converted from radiation to electricity in the photovoltaic cells?

But what about the miscellaneous heating that would occur to photovoltaic cells in a high level radiation field outside an order reversing field of a Refresher? The heating process, though not necessarily damaging and aging, also occurs as an ordered process in the second law of thermodynamics. If the order to disorder reversed field of the Refresher were added, the photovoltaic cells would be cooled down. Heating would not occur in the field. What would happen to the power ordinarily lost to heating? Would not it be added to the power converted from radiation to electricity in the photovoltaic cells?

But what about the gamma photons that would not age the photovoltaic cells or heat them, but would pass through them without affecting them? What if the Refresher field were added, what would then take place? The next question can resolve this question. Is the shielding loss included in the order to disorder arrows in the reaction equations? Yes. Then with the addition of the Refresher field, the elusive photons would return or never penetrate the photovoltaic cells. What would happen to that power? Would not it be added to the power converted from radiation to electricity in the photovoltaic cells? This result is the hardest to take. We need experiment to settle this. If this paragraph were not true, we would expect it would take layers upon layers—many feet of photovoltaic cells piled on top of each other to stop the gamma photons. But if this paragraph is true, then gamma rays as well as sunlight could be stopped by a single layer of photovoltaic cells in the order to disorder in the second law of thermodynamics reverser of the Refresher. In the reversed field, the photovoltaic cells should be 100% efficient. An EFP Reactor must be built and operate in the field of a Refresher.

While an individual photovoltaic cell may be 100% efficient, it would not be possible to cover every spot around the reactor with photovoltaic cells. But it should be possible to achieve 60% to on the order of 100% efficiency—enough for the source to be self-sustaining and an energy source.

9. What about Radioactive Wastes?

As we now experience the second law of thermodynamics, neutrons + products $\rightarrow$ neutron activation products. Reverse that and activation products become deactivated and neutrons are given off. Another reaction involving neutrons: $n \rightarrow p + e + $ anti v_e. Reverse that and neutrons are produced. In the field of the Refresher 1, neutrons appear stable. Also in the field, radioisotopes are all backed out of existence. As long as the Refresher 1 field is on, the EFP Reactor will be radioactive waste free. Each annihilation of a proton or neutron on the inside surface of the collision chamber of accelerator leaves behind a new radioisotope. But with the Refresher, these do not decay with decay heat. Instead, these have electron capture and alpha particle capture and un-decay cooling—and with the Refresher operating, maybe not even that, because it is a form of aging. Maybe there would be neither decay heat nor un-decay cooling. This distinction would have to be settled by experiment.

References

[1] SUMMARY TABLES OF PARTICLE PROPERTIES, January 1, 1998, Particle Data Group, as quoted by *CRC Handbook of Chemistry and Physics, 80th Edition* (Boca Raton: CRC Press, 1999), pp. **11**-1 to **11**-49.

[2] SDI: technology, survivability, and software (Diane Publishing Co., May, 1988), p. 140, NTIS order #PB88-236245.

[3] Mathias Deckers, Steam Turbine Blading Technology for Siemens, Germany, "CFX AIDS DESIGN OF WORLD'S MOST EFFICIENT STEAM TURBINE,"
http://www.ansys.com/assets/testimonials/siemens.pdf.

[4] Stephen Hawking, *A Brief History of Time*—From the Big Bang to Black Holes (New York: Bantam Books, 1988), pp. 144, 145.

[5] Robert B. Leighton, *Principles of Modern Physics* (New York: McGraw-Hill Book Company, Inc, 1959), p. 665.

[6] Francis Halzen, Alan D. Martin, *Quarks and Leptons* (New York: John Wiley & Sons, 1984), p. 107.

[7] David S. Saxon, *Elementary Quantum Mechanics* (San Francisco: Holden-Day, 1968), p. 386.

[8] Gordon L. Ziegler, *Electrino Physics* (Lacey, Washington: Electrino Energy, 2010), Chapter 16, http://.benevolententerprises.org/, A much better up-to-date *Electrino Physics* Draft 2 is available at createspace.com, amazon.com, barnsandnoble.com and orderable at 2,500 book stores.

Chapter 13

Project Description

A. Introduction

The Project Description is the micro-designing, construction, testing, and operating in ever expanding footprint coverage of the Refresher on earth and then on the sun, as the public demands and the governments permit; and the mass production and proliferating of the Clean Energy Sources around the world rapidly. Our objectives for this equipment are to make electricity free around the world, to abolish sickness, pain and death around the world quickly, to remove entirely all medical costs and funeral costs from the people by treating them and healing them for not one cent in fees. Other objectives are to halt and reverse global warming and to tame the sun by backing it up several thousand years through the Refresher technology.

B. Sketches

When the micro-design phase is done, we will have complete blueprints for everything for the Refresher as well as the Clean Energy Source. The only illustrations we have so far are crude sketches:

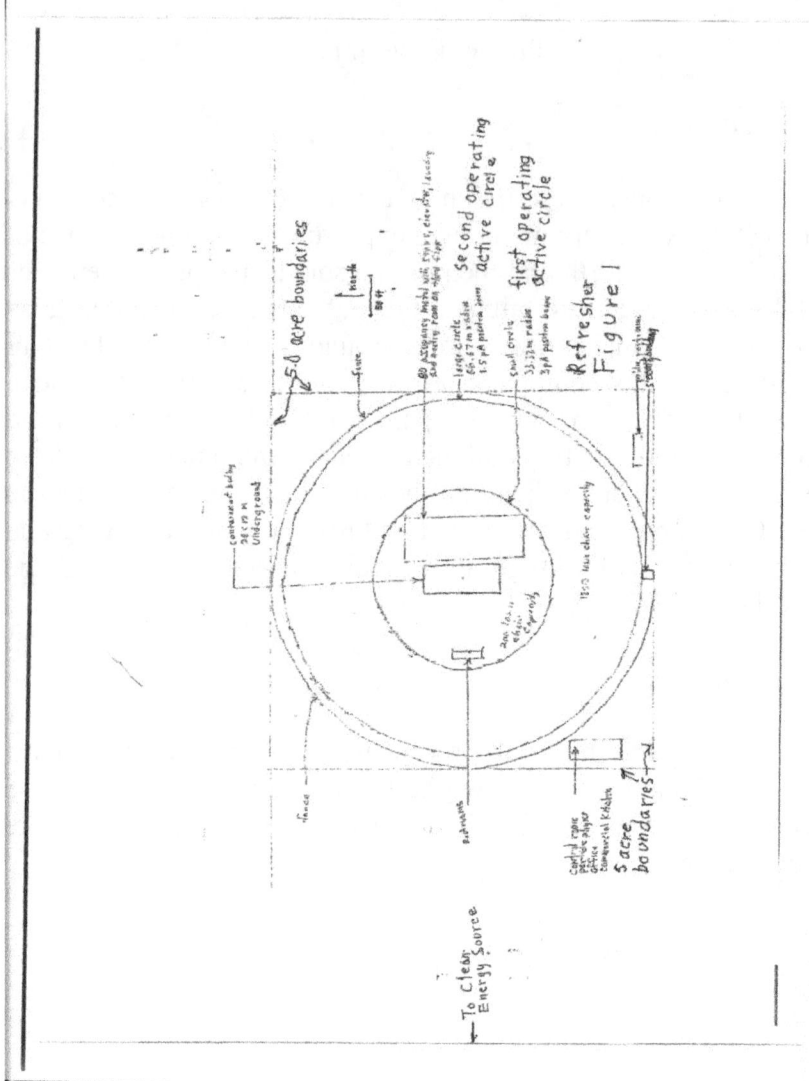

Refresher Figure 1

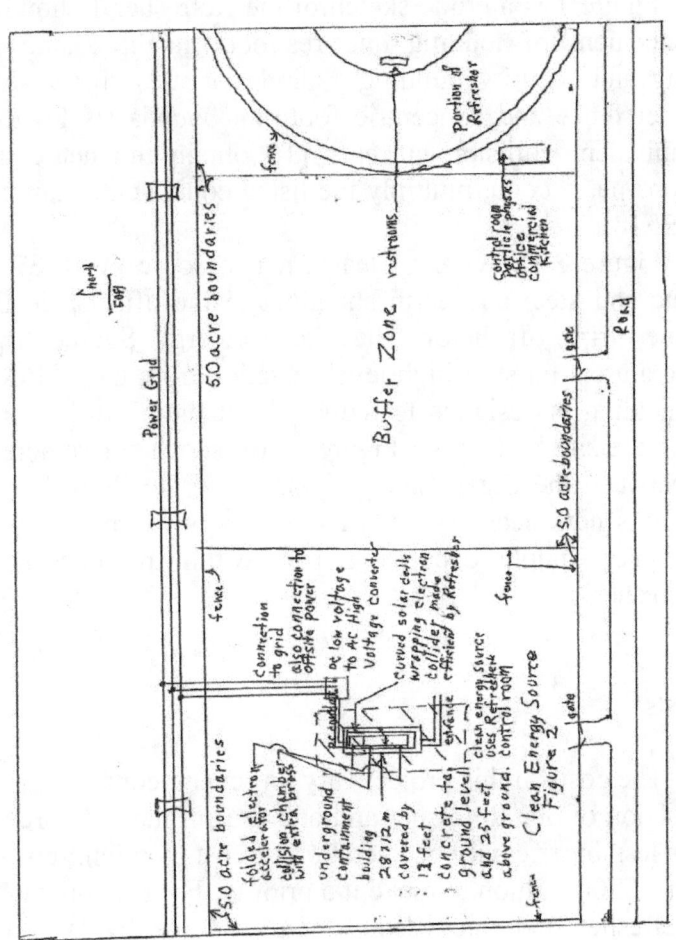

Clean Energy Source
Figure 2

Figure 1 is a crude sketch of the Refresher 1 showing also the placement of lodging and restroom facilities and control building and security building. Circles in this figure show the perimeter fence and concentric footprint boundaries for different net equivalent collision currents [To obtain the net equivalent collisions per second, multiply the listed equivalent beam currents by 6.25E18.].

Figure 2 shows the Clean Energy Source with 37 feet of concrete and steel and earth shielding in the off chance that the Refresher trips off before the Clean Energy Source trips off, causing a great burst of high level radiation from the EFP Reactor. The shielding is designed to reduce the radiation to less than 2.0 mr/hr at the site boundary. Figure 2 also shows a five acre buffer zone between the Clean Energy Source and the Refresher, which probably is not necessary, but is shown here for maximum safety to the general public occupying space within the fence boundary for Refresher 1.

C. Costs

The costs of this project vary on several conditions. For the construction of one Refresher and one Clean Energy Source in one year or less by rf cavity acceleration, it costs a minimum of $130 million—$200 million to make it a priority. For the construction of one Refresher and one Clean Energy Source by laser particle acceleration, it costs an additional $10 million (we should build both an rf cavity and laser style Refresher and Clean Energy Source). The total is $140 million.

Investments into Electrino Group Inc will be repaid five fold after the first year of operation. The first investor that finances the total $140 million will receive a 20% share of the corporation, and a paid up license to own, build, and market world-wide Clean Energy Sources in perpetuity. The costs of the design and construction by laser acceleration of particles for the Refresher is $5 million, and for the Clean Energy Source is another $5 million. The non-construction costs of the Refresher and Clean Energy Source are as listed below:

Additional costs approximately

Old debts from personal financing of mission—credit cards	$ 44,000
Old debts from the failure of Coherent Electron Source, LLC	772,000
Tying up of the property lease	3,000,000
EFSEC fee	100,000
Zoning, property and licensing fees	1,000,000
Developing property	2,000,000
Construction of containment buildings and control room	2,000,000
Construction of other buildings on campus	8,000,000
Architect/Engineer	1,000,000
Micro-designing of Refresher and Clean Energy Source	2,000,000
Personnel wages	5,000,000
Working capital, contingency	5,084,000
Total additional costs distribution	$30,000,000

If we can find suitable property with existing buildings on it, we may save up to $10 million dollars of this cost estimate. We have already found the ideal property if we can tie up the lease!

D. Emergency Situations

The situation from the meltdown and Fukushima Daiichi nuclear reactor tsunami disaster in Japan is getting worse and worse, already having high level radiation outside the containment buildings giving a fatal dose in 48 minutes, not permitting construction equipment in to drill holes for the pouring of concrete barriers to wall in the reactors to stop the migration of the high level radioactive materials from reaching the sea. There is no play book plan of what to do next. But if something is not done, the situation will become even worse. Worldwide contamination from this one source may soon doom the earth's populations to extinction through radiation induced cancers.

    The earth is going through a global warming crisis caused only in part by increasing greenhouse gasses caused by coal fired power plants and petroleum burning vehicles and homes. The far greater problem that effects global warming is that the sun is starting a nova sequence of getting brighter and hotter and hotter to peak much brighter and hotter than now, before getting less bright and hot and going out in darkness, freezing the earth and its inhabitants, dooming the populations of earth to extinction! This does not have to happen! The Refresher machine that the author has been inspired with could actually mitigate both the Fukushima Daiichi disaster and the nova on the sun as well as regenerate the earth.

Chapter 14

Refresher 1 Design Specifications

The design specifications for the rf cavity style Refresher will be identical or similar to those below. Final specifications will be available after the micro design phase.

Size of accelerator (folded)	20 m long by 3 m wide
Diameter of accelerator	100 mm (plus cooling channels)
Beam aperture	7 to 10 mm
Type of accelerator	Folded linear accelerator with pulsed klystron rf power supplies and S-band cavities (2856 MHz)
rf power supplies	Eight 35 to 50 MW pulsed klystrons
duty factor	0.1% (peak current 1000 times average current)
Average power	400 kW (20 kW per meter of accelerator)
klystron efficiency	~50%
total system power	800 kW

cooling water requirement for each 5 m section	5 to 10 gpm
cooling water required by each klystron	~ 5 gpm
cooling towers capacity	800 kW
Creation time total (if not super funded ($50 million))	3 years
Design time (beam dynamics, rf power systems, cooling, and computer control)	1 year
Fabrication and subassembly testing	18 months
Installation and commissioning	6 months
Creation time total (if super funded ($140 million)	1 year

Chapter 15

Trade Secrets

At first the author tried to keep the trade secrets proprietary, revealing them only to those that needed to know. But then my computer was hacked multiple times by different nations; and the contents of my hard drive and the trade secrets were available to different nations. Therefore the author chooses to make them free to everyone.

Trade Secrets for the Clean Energy Source and the Refresher 1
Specifications for the micro-design phase

Summary of Published Specifications

1) Accelerators for Clean Energy Source and Refresher 1 should run at 940 MeV.
2) A critical point published which could be forgotten or overlooked is that the electrons in clean energy source and positrons in refresher should be axial, focused as much as is humanly possible.
3) Beam currents for Clean Energy Source should run at 1 Amp, or as high below that as possible.
4) Operating beam currents for Refresher should be a maximum of 10 pico Amp with 13 or more lower beam currents by decades.
5) The electrons in the Clean Energy Source in one leg of the accelerators should be spin flipped before collision. The positrons in the Refresher in one leg of the accelerators should be spin flipped before collision, so have same sense in the center of mass frame.

6) For maximum efficiency, the linear accelerators should be folded with collisions in the center.

Electrino Fusion Technology Trade Secrets

The first trade secret I want to disclose is the relative power and priority of the forces of the Universe, including the two second law of thermodynamics "forces:"

1. The strong force of like attracting like, with aether motion faster than the speed of light and imaginary radii, is the strongest in the Universe. There are two separate formulations of this power— strong gravitational and strong electric. But these are equated in this Universe, so the strong electric can be used as a surrogate of the strong gravitational in most of our calculations. The strong gravitational and strong electric forces are like the Coulomb electric force with the inverse of the Fine Structure Constant $1/\alpha$ in the force equations: $F_{strong} = q_1 q_2 / 4\pi\varepsilon_0 \alpha R^2$ or $GM_1 M_2 / R^2$. It doesn't matter if it is two negatives or two positives for the charges. It is stronger than any other force.
2. The next most powerful force in the Universe is the meso-electric force of opposites attracting in slower than the speed of light aether motion in particle fine structures with real radii. $F_{meso\text{-}electric} = (q_1)(-q_2)/4\pi\varepsilon_0 \alpha r^2$. The author discovered this distinct force himself. It is a wonder that no other physicist discovered it. It was in plain sight all along in particle physics. This is separate than the Coulomb electric force and the strong force. This is the force that holds the dot • and the anti dot -• together in traveling orbit in the photon. This is the force that holds the plus spin electron (+ in chonomic structures) together in orbit with the positive charge uniton • in the neutron. This meso-electric force is the force between positive and negative octons ionized out of the vacuum, and is the ex-nihilo force of creating mass and energy out of nothing. It is a very powerful force, but it is weaker than the strong force.

3. The next most powerful "force" in the Universe is the positive (reversed) second law of thermodynamics ($\Delta E_O > 0$) "force." This "force" has the power to reverse the direction of all the interactions with which we are familiar—reverse aging adults to young adulthood, backing diseases out of existence, backing decay and pollution out of existence, resurrecting the dead, and regenerating the planet and the sun.
4. The second law of thermodynamics with which we are familiar ($\Delta E_O \leq 0$) is 1836+ times weaker than the positive (reversed) second law of thermodynamics. Though over 1836 times weaker than the reversed second law of thermodynamics "force," this is still a very powerful "force"—stronger than the Coulomb electric force, magnetism, electro-magnetic radiation (including gamma rays), the weak force, and ordinary gravity.

Adam and Eve, after their creation, existed in the positive ($\Delta E_O > 0$) second law of thermodynamics. But immediately after their moral fall, they existed in the default second law of thermodynamics state ($\Delta E_O \leq 0$), with which we are familiar. This is the state of adult aging, disease, decay, and death.

How do these priorities affect the processes in the Refresher 1?

1. Positrons of the proper amounts are injected in an accelerator with axial finely focused positrons (antimatter) and collided at 940 MeV each particle, with the positrons from one accelerator spin flipped before collision. This puts the positron anti-semions briefly in the same orbit attracted to each other by the strong force—the strongest force in the Universe. Once the conditions are met, there is no force in the Universe that can prevent the anti-semions from fusing to new unitons, the core particles of protons and neutrons. The core particles, the unitons, scavenge from the aether sea the residual necessary sub-particles to compose whole protons and neutrons. This process of creating new protons and neutrons reverses the order to disorder arrow in the second law of thermodynamics to the positive ($\Delta E_O > 0$) sense. The size of the footprint of the Refresher 1 is determined

by controlling the average beam currents in the accelerators by pulse gating the positrons at the spin flipper, which process is already published and is not a trade secret. That is all there is to the Refresher 1!
2. Within the active footprint of that Refresher-Regenerator, adult aging, disease, decay, and death are backed out of existence. Pain is abolished. This machine is worth everything, but, fortunately, it can be built for a few tens of millions of dollars. The Refresher makes the EFP Reactor (Electrino Fusion Power Reactor, alias Radioactive Waste-free Power Reactor, alias The Clean Energy Source) operate properly, safely, and efficiently!

How do the above force priorities work together in the EFP Reactor?
1. Before the EFP Reactor—the Clean Energy Source—is turned on, the Refresher 1 is turned on. For a split second it operates without the establishment of the reversed second law of thermodynamics, and there is normal default heating of the accelerator. But that is only for a split second. Then the constituent anti-semions in the positron beams fuse according to the collided beam intensities selected on the control panel. Then new protons and neutrons are created in the Refresher 1 collision chamber. New positive order energy is created in the collision chamber, which reverses the order to disorder arrow in the second law of thermodynamics within the footprint of the Refresher selected on the control panel. All the processes within the footprint of the Refresher now are totally under the control of the reversed second law of thermodynamics.
2. The first effect of that reversed second law is to back out of existence the heating of the Refresher accelerators, and preventing any additional accelerator heating to occur. The net result is to make the Refresher accelerators effective room temperature superconductive and 100% efficient, investing all of the applied energy into the acceleration of the positrons in the positron beams.

3. All of the Refresher 1 effects now occur in the Refresher footprint—reversing adult aging, backing disease, decay, and death out of existence in the selected footprint of the Refresher.
4. At the control panel of the Refresher, we make sure The EFP Reactor is in the footprint of the Refresher. Then we turn on the EFP Reactor and begin to ramp up its power levels. From the first, the reversed second law controls all the processes below instances of the strong force and the ex-nihilo meso-electric force. The electron beams in the EFP Reactor are made without heating—made super conductive and 100% efficient.
5. Highly focused 940 MeV spin flipped (on one accelerator leg) axial spin electrons collided in the EFP Reactor collision center at 1.0 Amp beams fuse electron semions to anti-unitons and anti-protons and anti-neutrons with or without the positive ($\Delta E_O > 0$) sense second law of thermodynamics. To achieve a safe efficient EFP Reactor (Clean Energy Source) we place the Source in the footprint of the Refresher-Regenerator. But don't worry, the powerful reversed second law of thermodynamics will not prevent the fusion of the negative charge semions to anti-unitons and anti-protons and anti-neutrons. The negative charged electron semions are the opposite charges as the positive positron anti-semions. But negative charges also fuse by the strong force—the strongest force in the universe. That force is stronger than the positive sense second law of thermodynamics, which cannot stop or prevent the fusion of the negatively charged electron semions.
6. The new anti-protons and anti-neutrons randomly drift away from the collision fusion spot and bump into the copper or brass collider housing and fuse to zero (annihilate) protons or neutrons in the housing by the powerful meso-electric force—opposite charges attracting. Again, the positive sense second law of thermodynamics cannot prevent or stop the matter and antimatter from annihilating and energizing ambient photons to oppositely directed annihilation gammas, because the fusing to zero (annihilation) process of the meso-electric force has more power

and priority than either the positive or negative sense second law of thermodynamics.
7. But the positive sense of the second law of thermodynamics is stronger, and has more priority, than the Coulomb electric force, magnetism, and electro-magnetic waves (included gamma rays). Now the gamma rays and the processes in the Refresher and the Clean Energy Source and in all the footprint of the reversed second law of thermodynamics come under the complete control of the positive sense of the second law of thermodynamics, which prevents gamma rays penetrating the copper or brass collider housing from aging or degrading the copper or brass housing of the accelerator and collider. The gamma rays go right through the copper or brass housing like super boson gravitons through the earth without colliding. They are forced to dump all their energy in concentric 100% efficient photovoltaic cells outside the collider housing. This feature of the operation of the Clean Energy Source is already published in *Radioactive Waste-free Power Reactor*, and is no longer a trade secret

Additional Trade Secrets

1) For balanced mix of anti-protons and anti-neutrons, electron collider should operate at 940 MeV. For balanced mix of protons and neutrons, positron collider should operate at 940 MeV. For anti-proton and protons only, run at 939 MeV. For shutting down colliders without turning off all power, reduce the accelerators to 938 MeV voltage.
2) Beam currents for the Refresher should be controllable by key lock switches on the console. The first switch should have two positions: testing; and decades. Under testing there should be two positions on another key lock switch: 3.0 picoamp beams; and 1.5 picoamp beams. Under decades on the first switch should come a rotary key lock switch with at least thirteen positions for thirteen different decades starting with 10 picoamps, then going to 1.0 picoamp, etc., down to 1.0E-22 Amps.

3) The decades are to be achieved in beam pulse technology with 1 peak for 10 picoamp; 1 peak + 10 peak size blank spaces for 1 picoamp; 1 peak + 100 peak size blank spaces for 100 femtoamps, etc. This would be easier to achieve than 1 peak and 9 spaces; 1 peak and 99 spaces; and 1 peak and 999 spaces, etc.
4) The pulse signals should control the spin flipper on the leg of one accelerator. The peak pulse should correspond to when the positron is spin flipped properly to make the two pairs of anti-semions to briefly be in the same orbit to fuse. The blank spaces between peaks should correspond to no spin flipping of the axial oriented positrons, making the beam shoot blanks. This would not be quite as difficult as colliding single particles. You would have whole pulses of the same size with many particles in them to collide.
5) The best news is that the reversed sense of the second law of thermodynamics will permit neither heating nor cooling of the apparatus controlled by the reversed sense of the second law of thermodynamics. This would apply not only to the particle residue after annihilations in the collision chamber, but to radioactive waste tanks at Hanford and all around the world in spent fuel tanks, as long as they were within the footprint of the Refresher. We need not fear extreme un-decay cooling freezing the tanks. It just simply would not occur in the footprints of the Refreshers!
6) The only thing to be concerned about is if the drifting anti-protons and anti-neutrons would always go in a precise direction, burning a hole in the collider housing. But that should not occur, for the drifting velocities of the anti-particles should be random.
7) The Clean Energy Source would automatically do self maintenance in the field of the Refresher. It would never need maintenance for the life of the plant—over 100 years.
8) With 940 MeV acceleration of the particles, electrino fusion should take place with the focus or with the lack of it of the beams.

9) When the accelerator is replaced after the 100 years, as long as the Refresher continues running, there will be no radiation hazard from the remaining copper, but it could be salvaged and recast with added copper into a new accelerator housing.
10) The operating Refresher will do continuous self-maintenance on all the supporting facility and supporting equipment, so that there would be no need to retire the facility at all but only rebuild the accelerators with recast accelerator housing, which could be prefab before the switch.
11) If the Refresher 1 footprint took in the whole world, or at least a whole country, the electrical grid would also be in the footprint of The Refresher. The inventor long had evidence that this would make the electrical grid be super conductive with ordinary lines. This would prevent the 60% power losses of the grid, and make the planned $0.03/kwh electricity rates the true total price of the generated electricity. The reason that theory was not published before was that that was just one more feature to disbelieve and jeopardize the project, but it can be told in trade secrets.
12) One Refresher can take care of the whole earth and thousands or millions of Clean Energy Sources. But one Refresher per Clean Energy Source would be useful in making a world-wide field with holes in it where fearful people could go to not be guinea pigs in the first operation of the order reversing field, whereas the convinced, desperate, and brave would not have to wait for long trials before treatment, but could be treated from the first. "Two are better than one; and a threefold cord is not quickly broken." It would be best to have at least three workable Refreshers in the world at all times, for if the Refresher active field went out, the system would revert to the pain and death we experience now. For safety and security, the Refresher could start with a footprint of the area in the radius of 10 meters, with the inventor, Gordon L, Ziegler, Kenneth Colvin and persons desperate to be relieved from pain and seizures to bear the initial risk of trying something new to be the first Guinea pigs (each

must sign a human use consent form). With no problems observed for 20 days, the footprint could be increased to the circle within a five square acre plot, and 10,000 persons with signed consent forms treated to observe if there are any negative side effects observed in the treatment of the first 10,000 people. If not, the governments should be petitioned to permit the enlarging of the footprint outside the fenced five acre test plot, so the public could freely expose themselves to the life-giving field without consent forms. Then the footprint could be enlarged a step at a time, as the public demands and the governments permit, until the whole earth is within the active footprint of the Refresher(s), and all people of the earth are pain free and are young adults again. Then the Refresher footprint could be enlarged to include the sun, so the sun could safely be backed up thousands of years, and tamed, halting solar induced global warming.

13) About the conversion of the gamma rays to electric power, we are to be prepared to make a hundred layers of concentric curved solar cells about the accelerators and collision chamber to approach 100% efficiency, but at first we are to try just three layers in such a way that all the cracks in between the cells are covered. It would be impossible to cover every square inch of space on the ends, and so we could never achieve 100.00% efficiency. But we can come very close to it. We want the photovoltaic panels to be curved symmetric, not flat boxes about the accelerators and collision chambers, for we do not want to introduce terms of cosine or sine into the efficiency calculations. How could the solar cells be 100% efficient to stop gamma rays, and not the copper housing to be 100% efficient to stop the gamma rays? Because the Refresher prevents several available pathways to the gamma rays (*see Radioactive Waste-free Power Reactors*) forcing the energy dump into the solar cells.

14) As a precautionary measure, make automatic shut off and manual shut off controls of the EFP Reactor if there is any threat to the shut off of the Refresher(s).

15) As for instructions for how to build the accelerators and inject beams into them, pay Jim Potter, Ph.D. of JPAW to do it and make micro designs for it. He knows how to do it. Don't try to do it yourself. Let him be at least your Technical Director. For laser particle accelerator Refresher or Clean Energy Source, Let Brad Sorensen be the Technical Director.

Gordon Ziegler June 23, 2014 5:42 AM PDT

Chapter 16

Trouble Shooting

1. No detectable reversed arrow to the second law of thermodynamics.
   A. Check and make sure the Refresher power switch is turned on.
   B. Check and be sure you have positron accelerating beams.
   C. Check and make sure the spins of the positrons in the beams are aligned axially.
   D. Check and make sure that the positrons from one beam are spin flipped before collision.
   E. Check and make sure two or more positrons are colliding. This may be difficult to achieve with low power beams. Make sure the positron beams are focused as much as humanly possible, and make sure the two beams collide. This also may be difficult to achieve. At first collide the beams at highest power, and then reduce the power as needed when the beams are proven to collide, and you want to calibrate the collided positrons to the desired level for the 1 picoamp collision rate baseline value (9.6 meter radius machine footprint). The boundary between the reversed order to disorder arrow of in the second law of thermodynamics and the non-reversed arrow should be abrupt and distinct.

2. Still no detected reversal of the order to disorder arrow in the proximity of the positron collision chamber.
   A. Check and make sure that the terminal energies of the positrons at collision are really 940 MeV. That may be the design

criteria, but the actual terminal energies may be off two or more MeV. If the terminal energies of the collided positrons are really 938 MeV or less, the system will not work at all! If the terminal energies of the colliding positrons are really 939 MeV, protons only will be produced in the collider. At 940 MeV, a balanced supply of protons and neutrons will be produced. At over 940 MeV, the system will still work, but will be less efficient.

How can the positron's energies be measured accurately? Let the beam go through a slit, then put the beam in a magnetic field to make it curve, then measure the radius of the curve, and calculate the mass-energy of the positrons from the measured curve radius, and the strength of the magnetic field. When you obtain the necessary intelligence, restore the collider with appropriate focus and collisions.

3. Still no detected order to disorder arrow reversal.

The system of second law reversal may actually be working microscopically, but unobserved. The effective collision current may be way too high to detect the effect. The system is counter intuitive. The higher the effective collision current, the smaller the radius of the footprint of the effect. And the lower the effective collision current, the larger the radius of the footprint of the effect. It would be difficult without timed gating to obtain effective beam collision currents small enough. The benchmark current for 9.6 meter radius of the machine footprint would be 1 picoamp (1×10^{-12} Amps). That may be difficult to achieve with ordinary electronics. But larger effective collision currents may reduce the effective machine footprints to microscopic values: A footprint of radius 9.6 micrometers (9.6×10^{-6} meters) would be achieved by an effective collision current of 1 micro amp (1×10^{-6} Amps), and a 9.6 nanometer footprint would be achieved by an effective collision current of 1 milliamp (1×10^{-3} Amps).

If the effective collision current for our benchmark footprint of 9.6 meters is too high, one beam could be defocused a little or a lot to change the collision cross section.

When the benchmark of 9.6 meters radius footprint for 1 picoamp effective collision current is achieved, you will want to have thirteen or more decades of larger footprints for smaller effective collision currents through an electronic timer gating machine, spacing pulses controlled at the electromagnetic spin flipper. If the spin flipper is energized by a pulse, the spin of the positron should be flipped to make the spins alike in the center of mass frame to enable positron collision and fusion to protons or neutrons. If the spin flipper is not energized by a pulse, the spins of the positrons will not be compatible for a fusion, and thus no protons or neutrons will be produced, and the order to disorder arrow in the second law of thermodynamics will not be reversed.

It would be easier to make an electronic gating by decades to make spaces between pulses 10, or 100, or 1000 pulse widths, etc., not 9, 99, or 999 pulse widths, etc. The various pulse timings should be set by an appropriate key lock rotary switch with 13 or more decade positions. The bench mark setting will be 1 picoamp for 9.6 meters radius of effective footprint. The high setting will be 1 collision fusion in 79 hours for rejuvenating the sun.

Chapter 17

Draft Environmental Impact Statement

1.4 Facility Environmental Impacts and Mitigations

The facility (Refresher 1) is a low beam current, high energy (over 1876 MeV in the Center of Mass Frame) positron-positron accelerator-collider. One environmental impact is that the system will emit positron, beta, and gamma radiation. Because the beam currents are very low (less than or equal to 10 pA each beam), the radiation levels (yet to be determined) of the accelerators and collider will be very low. The radiation levels will be mitigated further by twelve feet of earth shielding. The resultant residual radiation levels will be far below the federal de minimus level for radiation exposure to the general public (0.01 mSv per year to members of the public).

Whereas the general public will not be exposed to significant amounts of radiation by this facility, the general public may expose themselves to a field caused by the reversal of the order to disorder arrow in the second law of thermodynamics in the grounds of this facility. The field is beneficial to individuals and/or the planet in every way, so mitigations will not be necessary for the active field, except several relevant advisories will be posted for the visitors' comfort and pleasure. The following is the notice in English that will be posted around the periphery of the active area:

"NOTICE:
- **Welcome to the Eden-like field of Refresher 1. There is no charge for**

this blessing. But we accept donations.
- You may find that your aging is reversed to young adulthood in the active area of this field. You may also find all your diseases backed out of existence, so you find yourself in perfect health.
- However, there are certain things about this field you should be aware of before you enter the active area: the field will try to heal all your old wounds. All body pierced jewelry should be removed before entering.
- If you stay in the field long enough, the field will remove all tattoos. (God does not like tattoos [Leviticus 19:28.].)
- The field of the Refresher may make you infertile (we are not sure of this yet). You may not conceive any more

babies after entering this field, as long as you stay in the field. But you will remain immortal as long as you stay in the active field of the Refresher. If you want to conceive more babies, you must stay out of this field, but then you will stay mortal and will eventually die. After thinking about it for awhile, most would choose to have immortality without more babies, but the choice is up to you. The Bible predicts a time when there will be no more "infant of days." Isaiah 65:20. If and when the Refresher field is extended worldwide, this text may come to pass.
- The field also will attempt to resurrect dead animals and the leaves, stems, and roots of vegetables. To avoid the discomfort of animals

and vegetables re-organizing in your stomach, it is recommended that visitors eat only fruits, nuts, grains, legumes, and fruit-like vegetables in the active area and 72 hours before entering the active area. Food designed for Eden-like living will be provided for no charge for those within the active area, and for those out of the active area, waiting to enter."

[Refresher 1 is not only for good health. Particle detection equipment can be placed in the containment building of the accelerators to test for revolutionary physics data.]

2.0 Purpose and Need of Proposed Action

The purpose for this proposed action is three-fold: 1) detect high energy sub-particle fusion; 2) measure the efficiency of sub-particle fusion for power generation purposes; and 3) test the reversal of the order to disorder arrow in the second law of thermodynamics, thereby reversing aging, disease, and decay processes in a test area, and in larger and larger areas as the public demands and governments permit.

The need of the proposed action is also three-fold: 1) high energy sub-particle fusion is needed to simplify, give added symmetry and an additional layer of orbits to particle structure and increase parsimony to the science; 2) e^- e^+ collisions have 1.602×10^{-19} efficiency. But e^+ e^+ and e^- e^- collisions have a theoretical efficiency of 1.0 in the new model at over 1876 MeV in the Center of Mass Frame, due to the magnetic and weak forces making the particles smart bombs with respect to the strong force. That is a big difference of efficiency. Should all the efficiencies be 1.602×10^{-19}? If so, a new breed of power generators would not be possible which would be very helpful. Our facility would be able to measure the efficiency in more than one way. 3) Every sickness, every pain, and every death on this planet is a need for reversing the order to disorder arrow in the second law of thermodynamics, which our proposed Refresher 1 Facility might be able to do in a small test area, in larger and larger areas if the public demanded and governments permit.

3.0 Significant Issues or Sensitive Receptors

The Refresher 1, reversing the order to disorder arrow in the second law of thermodynamics, does many things that seem too good to be true. Thus the reviewer of these things may reject this project out of hand. But all these things happen naturally when that arrow of time is reversed. Notice a possible sequence under the existing second law of thermodynamics:

Healthy young adult → aging → cancer → death → cremation → scattering of ashes.

What if the arrows were reversed—all of the arrows? Well, not only could one be healed from cancer, made a healthy young adult again, but scattered ashes could gather themselves again, a person could be un-cremated, come to life again, be healed of cancer, made young again, and made a healthy young adult again, which is the maximum state of order in human beings. The system goes to and is stable at the maximum state of order with the reversed order to disorder arrow. But notice in all this, it is order that is increased, not the clock that is turned backward. The clock continues forward. Sometimes it is difficult to keep that distinction in mind in imagining second law reversals.

Let us imagine some second law reversals: A fallen down barn with weathered wood and rusty nails falls back upright, ceases to sag, and becomes a new barn again with new wood and new nails. A burned out forest un-burns. The burned to death forest creatures come alive again and are restored to health. In the battle field, explosives refuse to explode. Killings stop. Exploded ordinance un-explodes, restoring all that was lost. Slain soldiers are resurrected. Maimed soldiers are made whole again. There is no more pain or sorrow. Enemies are grateful for the blessings of life, and are reconciled. Generation after generation is resurrected, and the foundations of many generations are raised up. Such possible scenarios are endless. They all seem too good to be true, but they would all be possible if only the order to disorder arrow in the second law of thermodynamics could be reversed in a controllable area. That may be the case with the proposed Refresher 1. In view of the incredible blessings it may offer to us, we must permit the Refresher 1 to be built and tested.

What will be impacted the most? Those who have been dead the longest, whose remains are the most scattered. It would take the most machine time to resurrect them. It would take over 3.26 years machine time to resurrect those dead 6,000 years.

What are the most important impacts? The elimination of pain, suffering, disease, decay, and death. The Refresher 1 will be designed to do this in a small test area, in larger and larger areas as the public demands and governments permit, and for the entire earth at once, and even for the sun, when all the governments of the world OK it. [The planet would not be over crowded. See Section 1.4.]

It would be desirable to build one or two more Refreshers as backups when there is an outage (but there would be no outage).

4.0 Alternative Development

Are there better ways to meet the Purpose & Need? No. The only alternative the human race has had for thousands of years is the religious promise that God would come in the end of the world and do this for the saved few, whereas the lost great majority would perish eternally. The Refresher 1 promises to save everybody, without the loss of one! There is no scientific or religious alternative to this that can offer such a positive promise. And we do not have to wait in faith for centuries or thousands of years. We can have it as soon as we can build and test the device—in one year or less, or immediately when Jesus brings His working models with Him when He comes soon.

5.0 Affected Environment – Existing Conditions

The environment and wildlife habitats are increasingly stressed, depleted and threatened by increasing and intensifying natural disasters and over harvesting by man. Hundreds of species are going extinct. The planet is experiencing the dangers and loss of habitat by global warming. The environment still supports life, but time is apparently running out.

The Refresher 1 can operate in a small test area, but it is designed to operate globally. No baseline environmental data is provided for its small test area, but every large facility in the world, such as nuclear reactors, can provide current environmental data of their areas, which can be baseline environmental data for Refresher 1 gone global.

6.0 Environmental Consequences
Prediction of Environmental Impacts

The Environmental Consequences of the field of Refresher 1 have already begun to be told in the Notice posted and given out to visitors entering the field, and in section **3.0 Significant Issues or Sensitive Receptors**. It is impossible to predict every occurrence of law reversal that would be experienced. However additional typical cases that would be representative are presented here for both positive and negative impacts: Aging concrete would become like new concrete again. Houses with blistered paint on siding boards would not only have the paint become like new, but the underlying boards become like new—and there remain stable. Oxidized paint on cars would un-oxidize and become like new. The worn parts would un-wear. The engines would not operate backwards (that would be the backing up of the clocks, not the reversal of the order to disorder arrow).

There are considerations on a global scale that would not be a problem in a small test area. Planes and rockets would fly forward, but would be reversed in aging. Their parts would not wear out or fail. If the decision were made to go global, earthquakes would happen backwards. But instead of causing serious damage, the rubble would be convulsed and fused back into the original beauty of the buildings.

If things went to maximum order, people, animals, and vegetation would glow with coherent light suitable for individual flight like angels, not needing rockets or planes. There would be a bonanza of energy efficiency and availability. With so much light, the sun would not be needed.

7.0 Mitigation – What Will Be Done to Reduce or Prevent Impacts?

The field of Refresher 1 has impacts at every turn. But most or all of them are highly positive, not negative. They describe a much better world. It is the opinion of the Principal Investigator, Gordon L. Ziegler, that there are no true negative impacts from the reversal of the order to disorder arrow in the second law of thermodynamics.

It would be very frightening to the non-initiated, however, to see faces glow when it was not expected. Thus the best mitigation of the impacts would be an educational program through the media explaining these things. That work starts here with this DEIS.

Chapter 18

Tests for the New Model of Physics

Significances of the Unified Particle Theory and the Unified Field Theory

1. Einstein tried for 30 years to discover the Unified Field Theory and failed. There is a Unified Field Theory in the author's *Electrino Physics, Draft* 2, Chapter 7, where there is the union of several pairs of forces. In addition, Chapter 6 of that Volume, as amplified in the remainder of that book, and *Advanced Electrino Physics, Draft 3; Predicting the* Masses, Volume 1, Introducing Chonomics; *The Physics of Genesis*; and the volume, *The Higgs et. al. Universe* by the author, contain a part of a Unified Particle Theory. It is more parsimonious than String Theory and Many Dimensional Theories (this theory requires only three space dimensions and time and mass and their imaginary complements). It is more parsimonious also than the Standard Model of Quarks and Leptons. The Standard Model currently requires 61 quarks, anti-quarks, leptons, and anti-leptons and other particles to make up all known light and matter, but not gravitons. By contrast, it takes only one copy of an octon-anti-octon pair, in the Electrino Unified Theory, to create all copies of all possible particles in the Universe through different ionization states, fusion states, and combination states. One cannot get more parsimonious than that.

The Standard Model has seven particle structures for each eight particles. The first and the last structures are the same. But the masses are not the same! In the Standard Model, all the particles are not unique. But they are in the Electrino Theory.

One beautiful feature of the Electrino Unified Particle Theory and Associated Unified Field Theory is that in Chapter 12 of the book *The Higgs et. al. Universe* is a road map for the calculations of the masses of any "elementary" particle. In chapter 21 is seen that the masses of the particles calculated so far are within two or more place accuracy of the measured values. Every such two or more place accuracy of the calculated values compared to the measured value is a successful test of the physics theory. But the theory does not just test previously measured values—it makes predictions where there are yet no measured values. If the masses of particles can be calculated accurately, so too can their radii, their magnetic moments, and other particle parameters. As the constituents of the boson aether are better understood, the half-lives and lifetimes, also, of particles can be calculated. Thousands of measured physical constants could be reduced to just a few fundamental defined constants. There could come a Unified Constant Theory.

2. Revolution in Origin Theories. One thing the Electrino Unified Particle Theory does, it revolutionizes our theories of the origin of the Universe. Instead of needing to start with a Big Bang, the Universe could be created a step at a time from controlled octons. Instead of the interstellar red shift calculating a time for the Big Bang through the Doppler Effect, the Unified Theories present new mechanisms for interstellar red shifting—making the Universe much older than it is now thought to be. (See *Electrino Physics, Draft 2,* Chapter 14.) A Big Bang should make equal quantities of matter and anti-matter, which is not the case in the Universe. The Unified Particle Theory shows how to make more matter than antimatter in the Universe. (See *Electrino Physics, Draft 2,* Chapter 12 and *The Physics of Genesis*, Chapter 3 by the author.)

3. New Powers. So many have sought the Unified Field Theory for so long, that someone questioned if the theory would yield any new powers to man, or only be some quaint useless mathematical curiosity. The author is here to tell that this theory indeed does yield new powers and abilities to man.

This Theory shows how to reverse the order to disorder arrow in the second law of thermodynamics, thereby giving interesting properties to space—healing and regeneration. (See *Electrino Physics, Draft 2,* Chapter 16 and *Child Science* by the author.)

This Theory shows how to convert matter into antimatter, or vice versa, making anti-matter rockets possible, making new kinds of power reactors possible that can annihilate matter for power, and to clean up and safely dispose of high level radioactive wastes and toxic chemicals. This theory points in the direction of how to make inertia-less craft that would not have a speed of light barrier, but could hurtle through space at almost infinite speeds. (See *Electrino Physics, Draft 2,* Chapter 5.)

In Chapter 9 of *The Higgs et. al. Universe* is a listing of many lepton decay schemes. The science now is so advanced that each particle in each decay scheme has shown for it its calculated mass! But there is a large room for further balancing decay schemes of mesons and baryons.

4. Calculated masses of elementary particles may be the greatest tests for the Grand Unification Theory behind the Refresher 1 Theory. Already 22 particles have had their masses calculated and compared to the measured values (see below). The author is beginning a crash effort to calculate the masses of all the known remaining particles (about 200 more). The measured values of thousands of experimental tests are shown to correlate well with the calculated mass values by the author. This fulfills the great needs of the Standard Model of Physics scientists.

Let us first examine the mass predicting values of the electron family of particles, the pion family of particles, and the neutron family of particles.

Charged Lepton	Predicted Mass Ratio m/m_e	Measured Mass Ratio m/m_e	Measured/Predicted unitless
Electron e_0	1.000 000 000	1.000 000 000	1.000 000 000
Muon e_1	206.794 822 479	206.768 2843(52)	0.999 871 669
Tauon e_2	3 420.101 469	3 477.15	1.016 803 62

Pion family member	Calculated mass MeV	Measured mass MeV	Measured/Predicted unitless
Pion π_1	139.956	139.570	0.997 241 990
Kaon π_2	496.554	493.677	0.994 206 068
D-on π_3	1877.78	1869.62	0.995 654 443

Particle (m/m_e)	Predicted Mass Ratio (m/m_e)	Measured Mass Ratio (m/m_e)	Measured/Predicted unitless
n	1,828.202 115	1,838.683 66	1.005 733 253
Λ	2,032.495 772	2,183.337	1.074 214 781
Σ^0	2,288.490 108	2,333.9	1.019 842 730
Λ_b^0	10,618.179 61	10,996	1.035 582 407

What can we determine from the above data? The errors are not consistently increasing or decreasing along a consistent trend in any particle family. The calculated mass ratios and calculated masses all go in consistent trends, but the measured values do not go in consistent trends. The error factors are almost as much below 1.0 as above 1.0. This is more probably measurement errors than calculated errors. In the neutron family, the error ratios **do not** consistently increase for higher mass particles. The Λ particle has the greatest error ratio in the neutron family, and higher mass particles in the neutron family have lower error ratios. This is contrary to my previous estimates of errors in higher mass particles. Even in higher mass particles, there is at least two place accuracy of the calculations, which is pretty good considering they all depend on an in-exact measured Fine Structure Constant α.

In summary, the Neils Bohr method of calculating the masses of the particles is almost as accurate for the higher mass particles as the lower mass particles. I then can continue on in the particle mass calculations for the mesons and baryons like I have done in the books, *Electrino Physics,* Draft 2; *Advanced Electrino Physics,* Draft 3; and *Predicting the Masses, Volume 1, Introducing Chonomics;* and the book, *The Higgs et. al. Universe.* And I can try to make the calculations computerized in an Excel Spreadsheet Program. For that suggestion I thank Dr. Brad Sorensen.

For decades and billions of dollars, particle science has been on a quest to discover the Higgs boson and learn how to predict the masses of all the "elementary" particles. The mainline science community has not had a way to theorize and predict the masses of the Higgs bosons (H^0, H^+, and H^-), the Standard Model of Physics notwithstanding. Accordingly they have experimentally searched for heavier and heavier bosons. They have found one new boson at approximately 125 GeV. Is this the Higgs boson? According to a back of the envelope calculation in the Preface of *The Higgs et. al. Universe* book, the answer is no! The Higgs boson masses are calculated exactly to be much more massive imaginary values than the 125 GeV boson—on the Planck scale.

But much to the surprise of the author, the Higgs bosons are not only shown to exist, but to be the magic links between the mainline Standard Model of Physics and the author's Grand Unification Theory, The Electrino Fusion Model of Elementary Particles. The Higgs bosons are in everything in the Universe! H^+ and H^- orbit around each other and travel together at the speed of light in every ordinary photon in the Universe! Every baryon in the Universe has the H^+ particle as its core particle! And every anti-baryon has the H^- particle at its core particle! And every other particle in the Universe has ½, ¼, and ⅛ Higgs bosons in them.

In Chapter 12 of *The Higgs et. al. Universe* book is brought to light a roadmap for the calculation of the masses of every particle from first principles. Each different particle calculates its own mass from first principles from their unique radii of orbits. Furthermore straight line traveling zero mass electrinos can take on mass when injected into orbits. It is a way to make matter and energy out of nothing (ex-nihilo)! We could do this ourselves! The conservation of energy is only half correct. Energy cannot ever be destroyed. It is eternal. But mass and energy can be created out of nothing, expanding the Universe, which is a good thing! We need to keep the Universe expanding faster and faster, for, if the Universe ever collapsed on itself, all the joy and happiness and past achievements would have an end and be a failure!

The road map to calculating the masses of particles heavily depends on the science of chonomics derived in *The Higgs et. al. Universe*, chapters 3-10. Also, in chapter 21 we see that we can have two place accuracy or more by the Neils Bohr method of mass calculations even in the calculation of massive particles, when compared to measured values, which is pretty good, and the calculated values follow the trends better than the measured values, and thus the calculated values appear to be more correct than the measured values without any Somerfield ellipses or relativistic correction factors. That is a great tribute to the exactness of the measuring masses of the whole Peter Higgs et. al. scientific community.